W9-BER-291

# The GOODYEAR STORY

# The GOODYEAR STORY

by Maurice O'Reilly
edited by James T. Keating

A Benjamin Company Book

Managing Editor: Virginia Schomp
Art Director: Thomas C. Brecklin



Library of Congress Cataloging in Publication Data
O'Reilly, Maurice, 1912-
The Goodyear story.

Bibliography: p. 218
Includes index.
1. Goodyear Tire and Rubber Company. 2. Tire industry — United States — History. 3. Rubber industry and trade — United States — History. I. Keating, James T. II. Title.

HD9161. U54G526 1983 338.7'6782'0977136 82-73862

ISBN: 0-87502-116-6 Hardcover edition
0-87502-121-2 Softcover edition

Published by The Benjamin Company, Inc.
One Westchester Plaza
Elmsford, NY 10523

First printing: November 1983

0 9 8 7 6 5 4 3 2 1

*Front Cover Illustration:* Goodyear's Clock Tower. Built in 1915 as part of Goodyear's factory on Akron's Market Street, the Clock Tower symbolized the importance of time for early Goodyearites in those "driving days" of the burgeoning automotive industry.

# Contents

# Prologue

When Frank A. Seiberling founded the Goodyear Tire & Rubber Company in 1898, he was barely solvent and modestly acquainted with the rubber and tire business.

He was a 38-year-old entrepreneur whose family — headed by his father, John Frederick Seiberling — had started and successfully developed several business organizations in Akron, Ohio. They included flour milling, oatmeal production, banking, real estate, the Akron Academy of Music, a rubber company, and the manufacture of strawboard, mowers, and reapers.

What had taken years to build, though, took far less time to fall. Most of the Seiberling businesses were lost in the prolonged depression of the 1890s, which had started with the Panic of 1893. This four-year ebb in the economy resulted from careless railway financing, unsound U.S. banking practices, and an agricultural depression and was set off when both the Reading Railway and the National Cordage Co. failed early in 1893.

The liquidation of a Seiberling business called for Frank to visit Chicago in the spring of 1898, where he had a chance meeting with H. C. Nellis, an Ohio business acquaintance. Nellis was the secretary of F. Gray Co., of Piqua, owner of an East Akron strawboard plant that had been closed for four years.

The plant's seven-acre site, which Gray had bought from the Akron Woolen & Felt Co., had three structures on it: a small power plant and two dilapidated buildings facing each other on opposite banks of the Little Cuyahoga River.

Nellis asked Seiberling if he knew of any prospective buyers for the package, pointing out that even though $140,000 had been invested in land, buildings, and equipment, it could be bought for $50,000.

History has it that Seiberling said, "You'll be lucky to get half that." When Nellis responded that it might be disposed of for as little as $25,000, Seiberling's quick response was, "Cut that in two and I'll buy it myself."

Nellis offered the property for $13,500, and Seiberling bought it. The arrangement called for a $3,500 down payment and $2,500 yearly for the next four years.

Seiberling returned to Akron with doubts about his new acquisition: how to pay for it and what to do with it. On his first day home, though, he borrowed the down payment from his brother-in-law Lucius Miles, drew up a financing plan, and held the first of several conferences with his family on how to use the plant. His brother, Charles W. Seiberling, was especially helpful in setting goals.

A decision was made to enter the rubber business, and the new Seiberling enterprise was named the Goodyear Tire & Rubber Company, honoring Charles Goodyear, who died in 1860. Goodyear's 1839 discovery of the hot vulcanization of rubber had earned him the title of founder of the commercial rubber industry.

Frank was not jumping hopefully — and blindly — into an industry unfamiliar to him. He was thoroughly knowledgeable about the operations of Akron India Rubber Co., one of his father's earlier businesses, and had supervised its sale in 1897 to an East Coast business syndicate that was organizing the Rubber Goods Manufacturing Co.

His brother, Charles, had three years with Akron India. He had been its secretary for two years, remained with it after it was sold, and left only to join Frank and Goodyear.

The new company was started with capital stock of $100,000, of which $43,500 was paid in. Most of it came from the Seiberlings and their friends, including a few thousand dollars borrowed from Henry B. Manton, another brother-in-law and former treasurer of Akron India.

Henry Robinson, Manton's uncle and head of Whitmore, Robinson & Co. (later Robinson Clay Products Co.), subscribed for a portion of stock. A larger block went to David E. Hill and his son, George. The elder Hill was a close friend of the Seiberlings and also operated a clay products company plus several retail stores in East Akron.

The Goodyear company was incorporated on August 29, 1898. Charter officers were David Hill, president; George R. Hill, vice-president; Frank Seiberling, secretary and general manager; and Henry Manton, treasurer. These men and Frank's brother were also the company's first directors.

Although Frank headed the company from the outset, he did not become president until 1906. His starting annual salary in 1899 was $3,000; Charles became vice-president and purchasing agent in 1899 at $2,500 a year.

To begin operations, Goodyear needed additional capital, which because of the depressed economy was not always available. Frank was well-known for his dynamic personality, gutsy determination, and sales ability, but even his friends were skeptical about the new company. Nevertheless, he won over many Akronites who subscribed to $50,000 by the end of 1898, bringing the total to $93,500 — enough to start production.

In the summer and fall of that year, the old buildings on the Little Cuyahoga were renovated, second-hand equipment purchased and installed, and an operations staff hired. The first recorded payroll totaled $217.86, paid to a cleanup crew of carpenters and laborers during the renovation. The first purchase was a 60¢ accounts book. The first sale was 10¢ worth of kindling.

The plant was ready, the machines began to whir, and with the tooting of a factory whistle on November 21, 1898, 13 full-time employees began to manufacture Goodyear products.

A mere 18 years later, Goodyear was the largest tire company in the world, and in 1917 its sales exceeded $100 million. By 1920, an original investment of $10,000 in Seiberling's organization would have been worth $1 million, returning average annual dividends of $13,400.

Goodyear has retained its industry leadership ever since. In 1982, its sales were almost $9 billion, its employees numbered 132,000, and it operated 101 manufacturing facilities in 28 countries.

This is the story of how Goodyear grew — and why.

*Opposite, F. A. Seiberling, 1915. Left, strawboard factory on the banks of the Little Cuyahoga. Seiberling purchased the East Akron plant site for $13,500 in 1898 — and decided to go into the rubber business.*

*Charles Goodyear. Discoverer of vulcanization, the process that led to the commercial use of rubber, Goodyear died insolvent 38 years before the founding of the company that bears his name. This statue welcomes visitors to the World of Rubber museum in Akron.*

# Chapter I

# BEGETTING A GIANT

As the 19th century edged to a close, the Industrial Revolution was moving at full speed. A vibrant entrepreneurial spirit flashed throughout the nation, fanned by opportunity and increasingly high regard for business enterprise.

New U.S. corporations and their leaders were riding the waves of success to prominence abroad that was nearly as great as at home. Several of these corporations had their roots, or parts of them, in Akron, Ohio, a fast-growing city about 35 miles south of Cleveland. Diamond Match, American Tin Plate, Quaker Oats, American Sewer Pipe, and International Harvester were some of the flourishing firms rapidly — and indelibly — placing Akron on the map of international industrialism.

Until Frank A. Seiberling founded the Goodyear Tire and Rubber Co. there in 1898, though, the man for whom this fledgling business had been named was scarcely known to most Akronites. Charles Goodyear had been an inventor and a pioneer in the making of hardware. He died insolvent in 1860, but he will long be remembered by Akron and the world for one of the most widely used and significant developments of the 19th century.

While experimenting in January 1839 at his New Haven, Connecticut, home, Goodyear accidentally dropped some India rubber and sulphur on the hot surface of his kitchen stove. A simple mistake, but it paved the way to his discovery of the vulcanization of rubber, the key element leading to its commercial use.

A question lost in history is whether Charles ever set foot in Akron, Ohio.

By 1920, this city was the Rubber Capital of the World, a title earned after its population had tripled in the previous decade because of the rapid expansion of both the industry and the rubber products manufacturers in Akron. The Goodyear company had for four years been second to none in the production of tires, and almost since opening its doors in 1898, it had made clear to the world that Goodyear was and would continue to be an important constituent of the burgeoning American industrial scene.

## A Natural in Akron

The seed that begot Akron's leadership role had been planted in 1870 not by Goodyear, but by Benjamin Franklin Goodrich, a 29-year-old physician turned businessman. A constant supply of water, essential to rubber manufacturing; good transportation, provided by two railroads; an influential board of trade, with the promise of local support: these elements were primary considerations in Goodrich's decision to move his small rubber manufacturing operations to Akron.

His plant at Melrose, New York, near Albany, had struggled under the influence of heavy East Coast competition. Because no rubber companies had yet been established beyond the Allegheny Mountains, Goodrich sensed the opportunities of an expanding frontier and determined to move west.

He chanced on Akron through a promotional folder published by the Akron Board of Trade. He visited the city, liked it, and met with 21 leading citizens to outline the Goodrich company, its long-range goals, and why Akron was his choice as a new home.

## The Seiberling Connection

The group sent Colonel George T. Perkins, president of the Board of Trade, to Melrose to inspect the existing Goodrich facilities. Perkins liked what he saw, and his positive report helped influence 22 other Akron businessmen to join him in lending Goodrich $13,600 toward the move. With $20,000 more that Goodrich himself raised, he had enough capital to reestablish his business in Akron.

Among Goodrich's Akron backers was John F. Seiberling, a former president of the Board of Trade and Frank A. Seiberling's father.

In March 1871, Goodrich and his brother-in-law Harvey W. Tew of Jamestown, New York, opened Goodrich, Tew & Co. on Akron's Main Street, south of Exchange Street and near Lock One of the Ohio Canal. Its major product was fire hose, but the new company also produced belting, bottle stoppers, valves, rings for fruit jars, tips for billiard cues, and other rubber products.

Goodrich's venture had difficulty gaining momentum. The first five years produced little profit, and stockholders became restless. With financial help from Colonel Perkins, Goodrich bought them out — for $12,500. The company's capital stock was then $50,000.

In 1880, this was raised to $100,000. The organization was incorporated as the BF Goodrich Co., and its business steadily improved. As it did, Akron took its first steps toward world prominence as the heart of the tire and rubber industry.

Whether the Goodrich example influenced Seiberling to start the Goodyear company is unknown, though other powerful factors favoring the manufacture of rubber could not

have escaped him. The nation's bicycle boom had accelerated in the 1880s and peaked in the late '90s. The Automotive Age was beginning, and many U.S. industries were uncovering many new uses for rubber products.

No doubt Seiberling chose Akron to undertake a manufacturing endeavor because it was his hometown. His family was prominent and had enjoyed a great deal of business success there, even though the Panic of 1893 and the depression that followed had damaged Seiberling investments. He was well-known in the community, and if he needed financial backing, it would be easiest to find in Akron.

This growing Ohio city was a good choice for other reasons as well, especially as the industrial climate began to improve after the lean years of the mid-1890s. Akron's population of English, Irish, German, Dutch, and middle-European descendants provided a ready labor pool of skilled unemployed. More than 2,000 Akronites had lost their jobs in the depression, a heavy blow to a city of 40,000 residents, and in 1898 many were still seeking a good job and steady pay.

*Early Goodyear tires, including solid carriage tire (left), S-V carriage tires (back), gun carriage tire (front center), and auto tire with All-Weather Tread (right). The battle over carriage tire patents nearly scuttled the fledgling company, but a Federal Court ruling in 1902 vindicated F. A. Seiberling's stand against monopolistic licensing practices. The All-Weather Tread, with its distinctive diamond-shaped pattern, was developed by Goodyear in 1908 to answer the need created by the speedy new horseless carriage for a safe, skidless tire.*

Two soon-to-be automotive manufacturing centers were nearby: Detroit, 180 miles away, and Cleveland, a mere 35. Obviously, Seiberling saw a future in this infant industry, rapidly taking shape now that the concept of a motorized carriage had been proved practical.

Hot on the tracks of Charles E. and J. Frank Duryea's first American gasoline-powered automobile, introduced around 1892, came the Stanley-Steamer, Hertel, Haynes-Apperson, Oldsmobile, Autocar, Ford, Whitney, and a host of other varieties of autos. In 1908, 65,000 automobiles were produced in the United States, and the horse-drawn carriage was well down the road to history.

Goodyear's initial line of products aimed at the company's primary target: the high-volume markets for bicycle tires, most of which were pneumatic, and carriage tires, most of which were solid. Second-line products included horseshoe pads, rubber bands, sundries for druggists, and poker chips. The company recorded its first product sales — rubber sundries — on December 1, 1898, totaling $25.80. Sales for that entire first month in business amounted to $8,246.72. Goodyear produced its first automobile tire in 1901.

Competition in the tire industry was severe, especially in Akron. Goodrich, particularly strong, and the Diamond Rubber Co. were already established as industry leaders. Although the French manufacturer Michelin had first applied pneumatic tires to a motor vehicle, BF Goodrich manufactured the first pneumatic tires in the United States for this application. Made of a simple single-tube design, these tires were produced in 1896 for the Winton automobile, a Cleveland product. Single-tube tires dropped into disfavor, though, because they were difficult to repair; by about 1900, they were altogether discontinued for use on automobiles.

## Early Patent Problems

Goodyear was beset by many obstacles when it tackled the tire industry in 1898, especially because of restrictive private patents that inhibited the bicycle market.

When the bicycle surged into popularity in the 1880s, John Boyd Dunlop of Belfast, Ireland, obtained broad bicycle tire patents in England. His company then acquired other major patents, notably one for fastening the tire to the wheel by means of a wired-on method, which applied a pneumatic tire to a flat-base rim with detachable flange.

Dunlop's patents allowed him to develop substantial inroads into the pneumatic tire industry until a long-forgotten patent issued to Robert W. Thomson of England came to light. In 1845, Thomson had patented the principle of the pneumatic tire, but when he died in 1873, his patent quickly became obscure and lay forgotten for two decades. Its discovery, however, loosened Dunlop's hold on the industry, and tire manufacturers were freer to compete in an open sales arena.

Patents continued to plague Goodyear, though, even when Dunlop's influence had diminished. Owners of patents were under no legal obligation to grant a license to anyone who asked for it, and infringements — real or accused — were an occupational hazard for new companies trying to manufacture products. About a half-dozen companies dominated bicycle tire sales, including, besides Dunlop, two Chicago-based rubber manufacturers — Gormally & Jeffery (later a part of U.S. Rubber Co.) and Morgan & Wright — and Hartford Rubber of Connecticut. Goodyear had purchased various bicycle tire patent licenses, but because its aggressive sales operation sometimes threatened the patent holders, some licenses were revoked.

## Bicycle Sales Soar

The company's best sales channel for bicycle tires was through jobbers who sold tires for replacement on a competitive-price basis, therefore pitting manufacturers against one another. Goodyear did well in that market, building 4,500 bicycle tires in one day in early 1900. This quick and steady growth alarmed the Single Tube Automobile & Bicycle Tire Co., owners of the Tillinghast patent under which most Goodyear tires were being produced.

About a dozen other companies were making and selling tires under the Single license, which demanded a fixed sale price of $4.25 a pair for guaranteed tires, $2.76 a pair for nonguaranteed tires. Goodyear refused to meet that requirement, and Single revoked its license with the upstart company from Akron.

This move was a severe blow to Goodyear and nearly closed down an entire department.

That challenge was quickly met, though, by the development of a two-ply tire made with a strip of muslin between the plies to keep them separate until vulcanization. To reduce costs, Goodyear eventually replaced the muslin with tissue paper, actually toilet paper purchased from Michael O'Neil's department store on Akron's South Main Street. O'Neil's oldest son, William, founded the General Tire & Rubber Co. in Akron in 1915, naming his father its first president.

The two-ply bicycle tires sold well until a patent infringement suit was again brought against Goodyear by Rubber Tire Wheel Co. of Springfield, Ohio, which was headed by Edwin S. Kelly. As this lawsuit developed, the patent owners felt less and less confident in their position and eventually dropped all charges. This was a major victory for Goodyear, and its bicycle tire sales rolled on in their upward spiral.

Developments were not nearly as rosy for Goodyear in the carriage tire industry. Before 1890, tires for carriages were on the whole poorly designed and tended to separate from the wheels on bumps or sharp turns. Kelly had come up with a solution for that problem when he acquired an effective tire-fastening device from Arthur L. Grant. These two men formed a partnership in 1894 when they founded Rubber Tire Wheel.

With an eye on the expanding carriage tire market — and perhaps with another on the horseless carriage — Frank Seiberling set out after a license to use the Grant patent. The American Tire Co. of New York City was then manufacturing carriage tires without a license and being sued for it by Kelly and Grant.

The entire tire industry hoped that American Tire would win the case, but a decision rendered by the U.S. circuit court in Brooklyn on December 23, 1898, found in favor of the patent holder. Officially, then, only companies licensed by Kelly could produce Grant-type tires.

The Kelly-Grant organization sold the Grant patent soon thereafter for $1.2 million to a New York City syndicate organizing the Consolidated Tire Co. Consolidated made no tires, though. Its principal asset was the Grant patent under which it licensed other companies to manufacture and sell tires.

In 1899, Seiberling applied to Consolidated for a Grant license — and was refused.

## Battling a Monopoly

If the founding of Goodyear testified to Frank Seiberling's determination and self-confidence, his reaction to the license refusal confirmed it. Certain that the U.S. circuit court's decision against American Tire would not stand up under appeal, he decided to go the early route of American and manufacture carriage tires without a license.

His position, he said, was that he was willing and ready to pay royalties to those holding patents for the ingenuity, innovation, and hard work behind successful inventions. But to grant patent licenses to some and refuse them to others was unfair, monopolistic, and illegal. So too was the patent holders' dictation of prices, he said, for products made under their patents.

Goodyear's carriage tires differed slightly from tires patented by Consolidated — which had become standard in the industry — because of a flange that prevented dirt from squeezing between the rim and the tire. The company did apply for a patent, but continued to sell its new product, and at prices considerably below those of Consolidated. As a result, Goodyear's carriage tire business boomed, obviously affecting Consolidated's sales.

In defense, Consolidated agreed to give Goodyear tire production orders totaling $50,000 a month — if Goodyear would stop underpricing its tires. Seiberling consented.

But Consolidated turned down the first shipment of Goodyear tires it had purchased, claiming they failed to meet standards. Seiberling was convinced his tires had been rejected for competitive reasons, not quality, and that meant war. He set out to battle the monopoly.

Carriage tire production on the Little Cuyahoga was increased to the maximum, and sales soared. As expected, Consolidated sued for and won a cease and desist order. Seiberling appealed, and the court allowed Goodyear to produce tires under a bond of $1 million, a security that was finally reduced to $250,000 and posted by Frank's brother-in-law R. C. Penfield of Willoughby, Ohio.

The ensuing court case was the first of many early crises in the history of an American enterprise that in less than two decades would be the world's largest tire company.

Despite the long litigation that dragged on until late 1902, Goodyear sales soared. In 1899, the company's first full year of operation, they totaled a robust $508,597, generating profits of $34,621. In 1900, sales reached $1,035,921.

The courtroom wranglings, however, gave Goodyear only shallow prosperity, as all profits from the sale of carriage tires had to be placed in escrow. Until the court reached a decision, Goodyear's working capital came mainly from the sale of single-tube bicycle tires.

## Litchfield Arrives

In 1900, Goodyear hired its first technically trained tire man. Paul W. Litchfield joined the company on July 15 as factory superintendent. He was 25 years old and an 1896 graduate of the Massachusetts Institute of Technology (MIT). His starting annual salary was $2,500. Litchfield became interested in the rubber industry during his senior year in college and spent many weekends visiting rubber factories in New England. At graduation, he joined L. C. Chase & Co. of Boston, a manufacturer of bicycle tires. He left Chase in 1898 for the New York Belting & Packing Co. and in 1899 joined the International Automobile and Vehicle Tire Company as superintendent.

Litchfield had designed some tires at International for Locomobile and Waltham Orient automobiles and the Fifth Avenue bus line of Richard Croker, Jr., whose coaches were the first to use pneumatic tires. Frank Seiberling had heard of Litchfield in connection with the bus-line assignment and induced him to join Goodyear.

In his later years, Litchfield confessed he was not impressed by either the company's financial operations or its factory. He was interested in tire manufacturing, though, and immediately liked the Seiberling brothers, quickly developing an admiration for both the dynamic F. A. and the courtly, considerate C. W. As Litchfield wrote a half century later, he figured that at 25 he could afford to gamble a couple of years.

The first office of the man who would rise to be Goodyear's chief executive officer for three decades had barely enough space for a desk and a drawing board. But his responsibilities were extensive. Besides running the factory, Litchfield was the tire designer, rubber compounder, and head of personnel. He even came to be known as the company doctor because his new office had a medicine cabinet on the wall with a first-aid kit and — for restorative purposes — a bottle of whiskey. Almost a teetotaler, the young superintendent soon replaced the liquor with spirits of ammonia.

From the start, Litchfield won the respect of both the office staff and factory personnel because of his fair and humane treatment of employees. He made frequent tours of the entire factory to investigate operations and to get acquainted with the workers.

During one of these early visits, he chanced on a half-dozen men — including foremen — playing poker in an out-of-the-way corner of the plant. Displaying neither surprise nor chagrin, Litchfield said: "This is the last poker game you'll play here. I appreciate your interest in the game — in fact I share it — but it can't be played on company time. Instead, we'll play once a week at my home."

For many years thereafter a poker game for supervisors was a weekly event at the Litchfield home.

Seiberling and Litchfield were almost complete opposites. The founder of Goodyear was a short, daring, and mercurial opportunist of great persuasive power — a "natural-born salesman," according to his associates.

Litchfield, who built Goodyear into an industrial giant, was tall, dignified, and analytical. A man of strong convictions, he was dedicated to the improvement of Goodyear's position and the quality of its products. His main interests were engineering, design, research, production, and industrial relations.

The two ideally complemented each other and gave the young company a balance between sales and production, a Goodyear characteristic that has survived to this day.

## High Sales, Low Profits

Goodyear found itself facing a paradox soon after the turn of the century. Booming sales had built bicycle tire production to 400,000 in the 1901 fiscal year, but profits were thin. Further, the threat of losing the suit by Consolidated concerning the Grant patent hung over management like a heavy sword.

Nevertheless, the small group kept all operations moving at top speed in expectation of winning. Had they known of the world-changing significance of an event in Texas, their hopes for the future would have soared sky-high.

On January 8, 1901, the Texas oil industry was born when prospectors discovered Spindletop, the first big oil find in the United States, just a few miles from Port Arthur. That gusher and the others that followed in the same year spurred a rapid and sweeping development of the automobile. Further, with those gushers came the genesis of an industry hardly dreamed of by those early Goodyear managers, but an industry their company would dominate within 50 years: the manufacture of synthetic rubber. The first synthetic tire was still many years away, though, and Goodyear had more pressing issues at hand.

By the end of April 1902, the company's financial situation had become desperate due to continuing patent litigation. Suppliers were suspicious and made rubber shipments only on a cash-on-delivery basis. One monthly payroll was met through a quickly negotiated loan from an Akron department store, the Hower Co., and the fledgling Seiberling enterprise appeared doomed — to all but Frank and his die-hard associates.

On May 6, 1902, the federal court of appeals ruled that Goodyear had not infringed on the Grant patent, and the company's escrow funds were released. Goodyear employees celebrated en masse as work at the factory stopped completely to mark this significant victory, symbolically represented by a new broom attached to the factory's flagstaff.

That fall, Goodyear started construction of a new factory building with a floor area more than four times greater than the original plant.

*Above, Goodyear's first employment office. Opposite, early employees. Payroll in 1908 was 850 employees, and a new worker could expect to earn 15¢ an hour. In the driving days to come, payroll would mushroom along with sales and profits; in 1912, Goodyear employed over 6,800 people.*

Disaster had been averted, and Goodyear continued its drive toward Frank Seiberling's far-reaching dreams.

About a year after hiring Litchfield, Seiberling took aboard three men who would make great contributions to Goodyear's meteoric climb to the top. George M. Stadelman came on as manager of carriage tire sales; William C. State became master mechanic soon after joining the company; William Stephens took over as operator of the band-cutting machine.

The three S's — as they were sometimes known — were completely different types of personalities, but throughout their long Goodyear careers, each attested to Seiberling's skill in judging men.

Stadelman had seven years of sales experience with Morgan & Wright, but in appearance and demeanor he seemed an unlikely sales manager. Shy, gentle, compassionate, intellectual, and studious looking, he lacked the hucksterism usually associated with his contemporaries. In the 20 years he managed Goodyear's sales, in fact, he never made a speech.

But Stadelman was a genius in sales analysis and had great integrity, a firm character that inspired trust, and leadership qualities that constantly motivated his sales force. He moved steadily upward: general sales manager in 1906, vice-president in 1915, president in 1923.

State, a consulting engineer from Springfield, Ohio, had built an operable automobile when only 21 years old. On his way to join the manufacturer of Winton automobiles, he stopped in at Goodyear to visit a friend. Seiberling met him, quickly sized him up, and offered him a job on the spot. Impressed by the challenges and opportunities at Goodyear — and perhaps by the Seiberling eloquence — he accepted.

In 1909, State invented an automatic tire-building machine. Later he was involved in the planning and construction of Goodyear factories until the early '30s. In 1929, he supervised construction of the huge Goodyear Airdock in Akron, the largest building ever built without interior supports, as an assembly hangar for two big U.S. Navy dirigibles. This structure, still standing, is 1,200 feet long, 325 wide, and 211 high.

## 'Boys! Get Going.'

State personified the get-it-done attitude that characterized management in the early days. In the construction of new buildings particularly, his philosophy was, "We can't wait." His usual modus operandi was: contact a construction company (usually Hunking & Conkey of Cleveland, in which he and Litchfield had great confidence), ask for a building of certain proportions, and give the go-ahead with a "Boys! Get going." Blueprints, contracts, and refinements came later.

It was well-known that he nodded in approval as a Hunking & Conkey foreman sat on a mound alongside the site with a pile of rocks at his side, ready to throw one at any worker caught loafing or dropping a shovel. The roofs of the early buildings invariably were installed by Akers & Harpham of Akron, because, as State explained, "We know them, we trust them, and we're in a hurry."

Stephens quickly gained the affection and respect of the company's entire work force. A natural athlete, he was well-known in all of Akron's sports activities. High-spirited, an expert amateur boxer, and constantly at-the-ready to go fishing, Stephens joined Goodyear, it was said, only because he learned the company was organizing a baseball team.

Known by everyone as Bill or Steve, never Mister Stephens, he steadily moved up through the ranks and was made general superintendent in 1920, holding that position until his death in 1932. It was said that he personally knew more people at Goodyear than anyone else, and he was once described as "looking like a superintendent ought to look, a man who walked like he was going someplace."

With a nucleus of dedicated, hard-driving managers headed by F. A. and Litchfield, Goodyear moved rapidly ahead. By 1906, it had become the national leader in carriage tire sales, and management was already taking dead aim at the automobile market.

Litchfield, in fact, concentrated most of his energies on the development of automobile tires. He believed that to increase the company's two-percent share of this market, Goodyear had to revolutionize auto tire design and struggle free from the restrictions imposed by the Clincher Tire Association, which controlled all licensing to manufacture the clincher tire.

The clincher's bead, the section that holds a tire fast to the wheel rim, was made of rubber and had to be stretched over the rim. It was mounted and demounted with a crowbar, patience, and a lot of sweat, and it usually gave only about 2,500 miles of service.

*The clincher tire. Goodyear's straight-side, "the tire that made Goodyear," quickly made the clincher obsolete. The straight-side, introduced in 1905, was easier to mount and gave a smoother ride.*

Litchfield considered the auto industry to be on the verge of solving the engine problems that had been its major drawback, and he thought it would expand tremendously when easy-to-mount and longer-lasting tires were developed.

## 'Just What We're Looking For'

The first step toward the tire he envisioned was taken in 1900 when "Nip" Scott, an inventor from Cadiz, Ohio, came to F. A. and Litchfield with a machine that could braid wire. Scott had used it to produce bedsprings, but found it too expensive for volume production.

It was "just what we're looking for," Litchfield told Seiberling. With the braided wire cured into the bead by which the tire was held onto the rim, the tire locked tightly onto the wheel. All that was needed was a pair of flanges bolted to the wheel for the tire to ride in. A new tire had been born.

Because its sides could go straight down from tread to rim and need not curve in for locking over the rim — as the clincher did — this tire was later known as the straight-side tire. It had 10 percent more air space than the clincher, provided an easier ride, and quickly caught on with users.

To promote the new design, Goodyear ran a national advertisement in the *Saturday Evening Post* with this slogan: "The Best All Roll/They All Roll Best/On GOODYEAR Tires." The ad featured the Wingfoot trademark, which the company had adopted in August 1900.

The trademark had been Frank Seiberling's idea, inspired by a statue of Mercury on a stairway banister in his home. Mercury, the ancient Roman deity who was messenger for all the gods and a symbol of speed: what better symbol, F.A. thought, for a company in the business of transportation.

The straight-side tire gave Goodyear its first original equipment sales — 10 sets of 34″x 4″ tires to the manufacturer of Winton autos. Some of the tires were exported to England and gained wide public attention there when ordered by such notables as the Prince of Wales, Sir Thomas Lipton, and Lord Salisbury.

In 1902, Goodyear's new tires were entered in a 2,500-mile race in England. Litchfield, who wanted to see the results of this endurance test firsthand, took a month off and journeyed across the Atlantic. He came home disappointed — but determined. Goodyear had been outperformed by Dunlop and Michelin.

He went to the drawing board with a new idea, convinced the tires had failed not because they were too weak, but because they were too strong. The problem could be resolved by tires that would absorb the shocks of the road, not resist them.

Production techniques were changed to make the straight-side more flexible. Still dissatisfied with the way the tire was attached to the rim, Litchfield worked long and hard with F. A. and Scott to find a solution. Which they did.

## Think Tank to the Rescue

A metal locking ring was developed with ends that did not quite meet. This ring could be snapped into a groove in the rim to hold the flange in place. The tire could easily be removed by prying off the ring and releasing the flange.

Two obstacles remained. The rims would have to be manufactured, an added cost, and the automakers would balk at a production change-over from clincher tire rims to rims for a new and untried type of tire. Here were formidable deterrents indeed, but they were only minor frustrations in the long run.

"Nip" Scott invented a rim that changed the contour of the outer flange in such a way that it was reversed, turning in instead of out. A flange of this design, he reasoned, would fit both the clincher and straight-side tires. Seiberling immediately acquired patent rights, and the Universal Rim had been created.

In December 1903, Litchfield wrote out the first specifications for the Quick Detachable tire, as Goodyear publicly called its straight-side invention. Modifications over the months included a breaker strip of square-woven fabric between tread and tire carcass to help prevent separation caused by the strain and heat of increasing speeds, but this brought about only minimal improvement.

Goodyear's ever-ready think tank then came up with the rivet fabric, one with a more open weave that allowed the rubber to flow through the tread and carcass during vulcanization, anchoring them together. This new fabric not only reduced tread separation, but helped distribute road shock throughout the tire. No mill in the United States produced such a material, and until one was persuaded to, Goodyear hired a crew of women to fabricate it by pulling out cross threads by hand from square-woven fabric.

## A Promotion Gamble

The rivet fabric quickly became standard in all Goodyear tires, and in 1905, the U.S. public from coast to coast was told of the revolutionary and improved Quick Detachable tire. The introduction, with nationwide promotion and advertising, was spectacular for its time and called for full-page advertisements in leading national magazines, including, once again, the prestigious and widely circulated *Saturday Evening Post*.

Goodyear went all out to market the new tire, staking its reputation, and perhaps its future, on a promotional campaign that had taken nearly every dollar the company could muster. The tire industry was skeptical, but the gamble, as some called it, was successful.

The campaign gained helpful impetus from a five-day automobile endurance race in England. Buick and Reo, two leading U.S. automakers seeking a foothold in the British market, mounted the new Goodyear straight-side tires on their entries and made an excellent showing. The Reos finished the five-day grind without a single puncture, a remarkable accomplishment for that time.

Shortly thereafter, Bill Stephens arranged for the well-known racing driver Louis Chevrolet to use Goodyear's straight-side tires on his Buick-sponsored entry in a 500-mile race at the Indiana State Fairgrounds, predecessor of the Indianapolis 500 Speedway. Chevrolet failed to win, but his tire problems were minimal compared with those of his opponents. Much more important was the attention paid to Goodyear's new tire by the motor-racing aficionados.

As one car manufacturer after another gave the straight sides a try — and liked them — the tire industry gradually began to shift in Goodyear's favor. The Quick Detachable tire became original equipment on nearly all autos produced both in the United States and Europe, and Goodyear quickly became one of the half-dozen industry leaders. Daily production of the new tire steadily increased from 90 in 1905 to 900 in 1908.

## Battle in the Marketplace

Although Litchfield and his technical group had given Goodyear a solid competitive lead, the battle in the marketplace was fierce. Tires had to be sold aggressively, and the Goodyear sales force was more than ready to sell its new product.

George Stadelman's seven years with Morgan & Wright had given him both an in-depth knowledge of the national tire market and many acquaintances in the carriage and auto industries. He made it a policy to hire only salesmen who had experience in automotive industry sales, and they concentrated most of their sales efforts on the auto manufacturers, a strategy that moved Goodyear quickly into the top rank of original equipment suppliers for new cars.

In 1905, the sales force consisted of five branch managers — one each in Boston, New York, Chicago, St. Louis, and Cincinnati — and four traveling salesmen.

*The "No Rim-Cut," a straight-side multi-ply auto tire, is promoted in this 1911 advertisement in "Life" magazine.*

More than half of the Goodyear auto tires sold in 1906 to auto manufacturers were in two record orders: 1,639 sets to Reo and 489 sets to Buick. Other important original equipment customers that year included Autocar, Cadillac, Ford, Oldsmobile, Packard, Studebaker, and White, the cream of the crop among auto manufacturers.

## Growing with the Automakers

In 1907, Goodyear opened a branch in Detroit concerned mainly with auto industry sales. Little genius was needed to see that the auto industry was coming on strong and that Detroit was just as aggressively becoming the hub of activity. Goodyear's biggest single order that year was for 1,200 sets of tires to go on Henry Ford's wondrous new car, the Model T, which would be introduced in 1908.

In an era of stunts and other derring-do attractions, Goodyear hired Barney Oldfield, America's best-known racing driver, for some barnstorming autobatics to demonstrate the straight-side tire. He drove a measured mile in 51.8 seconds and established a 1907 record driving time of 17 hours and 17 minutes from San Francisco to Los Angeles.

By the end of 1909, Goodyear's original equipment orders made up a large share of its business, and in the January 1910 issue of the *Saturday Evening Post,* a daring double-page ad announced that more than a third of all automobiles produced that year would be equipped with Goodyear tires. The claim stood up; about 36 percent of cars sold in 1910 rolled off the line on Goodyears.

With the imaginative, promotion-conscious Seiberling at its helm, Goodyear was an active marketing organization from the beginning. In the fall of 1898, even before the company began production, it had exhibited tires at the St. Louis Carriage Builders Show. Goodyear's first recorded print advertisement appeared in *Motor Age* magazine on September 12, 1899.

National advertising in general publications, such as *Collier's, Harper's,* and the *Post,* was started in December 1900. Advertising expenditures for the 1900 fiscal year were $9,238. By 1903, they had reached $27,243.

During the company's first year in business, top management changes took place and continued throughout the first decade of the new century.

The first president, David Hill, resigned in 1899, reportedly concerned over Seiberling's dynamism and lofty ambitions. R. C. Penfield, who had provided financial support in the earliest troubled times, succeeded him. A year later, Lucius Miles, Frank's brother-in-law who had lent him the original down payment of $3,500 to buy the plant in 1898, took over that post. When Seiberling himself became president in 1906, he turned to the company's active managers to share that leadership with him. Litchfield and Stadelman were made directors.

In 1908, the hiring-in rate was 15¢ an hour, and an expert tire builder on piecework could make 30¢ an hour. The maximum hourly pay for a supervisor was 27.5¢. The company had 850 employees on its payroll: most of them lived close enough to walk to work and lunch at home.

The plant, which early that year had expanded into adjoining farmland after the success of the straight-side tire, had 5.5 acres of floor space. Horse-drawn wagons regularly delivered rubber and raw materials from the railroad station at Mill and Summit streets, returning loaded with tires for shipment to purchasers.

Automobile tire production jumped to 36,000 annually at Goodyear, and sales for the fiscal year that ended August 31, 1909, were recorded as $4,277,067, nearly doubling the previous year's volume. Keeping pace with the auto industry, Goodyear was moving — fast.

By that time, management had jelled under the creative guidance of Frank Seiberling who, when he saw an opportunity, wasted no time in acting on it. A promoter, inventor, risk-taker, and charismatic leader, he was well-known as the brains of Goodyear in its early days.

But if Frank was the brains, his brother Charles was the heart of the company. C. W. was charming, personable, and friendly, highly respected throughout Akron for his involvement and leadership in civic affairs.

Rounding out the team of four, which carried Goodyear to the number one ranking in the burgeoning tire industry during the years of 1908 to 1920, were Litchfield and Stadelman.

Litchfield established an early reputation as a production genius with superior executive ability, dedication, and clear-thinking insight into the industrial future. Stadelman was known as an astute analyst of business and market conditions.

A strong Goodyear spirit developed in the company during the formative years, a result of many events and influences. This spirit's strength and durability were probably most stimulated by the tribulations and successes of the company's shaky first decade and by the Seiberling-Litchfield conviction that "People Are Our Most Important Asset."

In recognition of that esprit de corps and to fortify it among the production force, in 1910, Litchfield formed the factory "Old Guard" of early Goodyearites, who had joined the company before 1900. It included F. A. Seiberling, George E. Swartz, Farran G. Hills, Albert B. Cunnington, Clara Bingham, A. J. Huguelet, Mary Higgins, Fred Colley, George Spaulding, Carl Klingenhagen, B. F. Weaver, Ed Viers, and Edward Hippensteal, Goodyear's first employee. Hippensteal had been hired several weeks before the company was incorporated to tidy up the old strawboard plant and replace its broken windows.

## 'Driving Days'

The U.S. automotive industry in 1908 was moving into the first phase of a production miracle that would profoundly affect the economy of the United States for the next 50 years and beyond. The mass production techniques developed by Henry Ford and his peers in the 20th century's first two decades would capture international attention and admiration and in the years to come would be copied on every continent in the world.

Goodyear went all out to keep pace with its main customer, the automobile. In the late 1940s, looking back to the company's early production expansions, Paul Litchfield said:

> The automotive industry was more than receptive to goods which would popularize travel by car; the public was eager to get tires which would run longer, give greater safety, greater freedom from interruptions and annoyance of road failures.
>
> With an improved tire in our hands, the need for saving time became imperative . . . If we could not produce the tires fast enough with existing buildings and machines, we must build new buildings, buy or design more or better machines.
>
> I have no hesitation in saying that in such a situation we paid more money for buildings and machinery than we would have done had we been willing to wait.

*A clincher tire is installed on a wooden rim.*

> Time was literally money to us, and the extra money we spent was spent deliberately, and with our eyes open, knowing that we were buying precious days and weeks and months which were worth much more than they had cost us. For those were the driving days of a driving industry.

In 1915, a clock tower was part of the first Market Street factory building to be built. For Goodyearites of those days, it symbolized the importance of time; for the next 30 years most Goodyear factories had clock towers. The original tower in Akron was called the Old Guard Tower for several decades because the Old Guard club of early employees held its first annual dinner and many of its meetings there.

Fred Colley regularly played the chimes that had been installed in the tower in September 1920. They were first officially heard just before midnight on New Year's Eve of that year as Old Guard members finished their reunion dinner.

Before late 1906, most of Goodyear's original equipment sales were to makers of low- and medium-priced cars. The auto industry leaders, such as Winton, Pierce-Arrow, Packard, and Locomobile, were regular customers of Goodrich, U.S. Rubber, and Diamond Rubber, all proponents of the clincher tire.

But on October 8 of that year, Goodyear's directors agreed to license other tire manufacturers to make and sell the Goodyear design of the straight-side tire and the Universal Rim. Licensees included International Automobile & Vehicle Tire Company, which soon was acquired by Michelin, the leading tire company from France. Michelin paid $186,000 in royalties to Goodyear over the years it manufactured under the agreement, part of the licensing revenues that helped swell Goodyear profits.

In 1907, Goodyear decided to sell the Universal Rim to manufacturers, dealers, and consumers with or without accompanying tire orders, a surprising change in policy. In effect, that sounded the death knell for the clincher, because everyone could then compare clincher and straight-side tires on a one-to-one basis. Goodyear quickly became known as a quality leader in the tire industry, and the straight-side tire was widely recognized as "the tire that made Goodyear."

In a 1938 speech to a group of Goodyear's west coast managers and sales representatives, Litchfield described management's attitude behind the all-out drive to promote the straight-side tire and the Universal Rim.

> We went out without any reputation and as small as we were to try to revolutionize the tire business at 20 percent higher cost than the standard tire. You can't do that unless you have something to back it up, something the public will realize is worth more than the difference in price.
>
> We threw the entire influence of the company behind that tire, challenging the world's standard tire . . . The results of our campaign were shown in our tire production increase, from 35,000 in 1908 to 1,084,000 in 1912 . . . Each year, we doubled the business of the preceding year, and the value of the Goodyear common stock increased 32 times in those five years.

*A Buick is fitted with a Goodyear tire. In 1906, Buick and Reo were the company's best original equipment customers.*

During the 1908-09 fiscal year, original equipment contracts were made with such large auto manufacturers as Cadillac, Cartercar, Jackson, Kissel, Marion, Mitchell, Oakland, and Overland. In all, tire sales were made to 115 individual carmakers.

## All-Weather Tread

Although Goodyear's early success is rightfully attributed to developments of the straight-side tire and the Universal Rim, a further innovation late in the company's first decade of operation made an important contribution to the strengthening of the Goodyear image.

Early automobile tires had smooth treads that provided little traction and made skidding a constant hazard. To eliminate that hazard, manufacturers developed treads designed for better traction. Some European tires of the time had steel studs imbedded in the tread, which tended to fall out during normal use.

Most U.S. nonskids had sculptured rubber treads, and some manufacturers had licenses for the Bailey tread, a band of rubber with round buttons vulcanized to the tire carcass, which enjoyed brief popularity. Increasing speeds and a horseless carriage that ventured more and more from the city to the farm, to rural roads of mud and ruts, compelled the tire industry to step up its efforts to find the skidless tire.

In 1908, Litchfield and his factory crew developed the diamond-shaped tread with four-way edges that resisted skids in any direction. Called the All-Weather Tread, it immediately became identified with Goodyear and for many years was mechanically better than any other nonskid tread on the market. For more than a half century, in fact, the distinctive diamond tread symbolized Goodyear around the world.

*General office entrance along Akron's Market Street. The Clock Tower, built in 1915 as part of the first Market Street factory building, symbolized the importance of time for Goodyearites. Over the next 30 years, nearly every Goodyear factory built had a clock tower.*

# Chapter II

# PROTECT OUR GOOD NAME

During the booming decade of 1910-19, Akron took shape as the Rubber Capital of the World. The automotive industry rolled cars and trucks off assembly lines in rapidly increasing numbers, and the tire and rubber industry grew accordingly. Partly by chance and partly by plan, its new leaders were headquartered in this northeast Ohio town that experienced a population growth from 69,069 in 1910 to more than 208,000 in 1920 — nearly all attributable to the surge in the rubber industry.

America's automakers, now entrenched in the Detroit area, were competing successfully abroad with European manufacturers, a direct result of production techniques that permitted low costs despite high wages. In turn, those high wages helped create a new domestic market of workers who could afford to buy the cars they helped to make. It was a healthy full circle of prosperity.

Car registrations in the United States increased from 8,000 in 1900 to 77,400 in 1905 and nearly 200,000 in 1908. The automobile, by this time capable of average speeds of 30 miles per hour, was still a sometimes annoying contraption, complicated to run and maintain, but it was here to stay as the key to a great growth industry.

Mainly because of the automobile, world rubber consumption grew from 40,000 long tons in 1900 to 297,500 in 1920.

Rubber trees were first cultivated in the Amazon valley of Brazil, and most crude rubber of the day came from there and Southeast Asia. Until shortly after 1900, Brazil controlled both the world supply and the prices of rubber.

In 1876, English botanist Henry Wickham smuggled 70,000 seeds of *Hevea brasiliensis* trees out of Brazil and took them to Kew, near London. More than 2,000 of these seeds were cultivated in hothouses there, and nearly all the plants were sent to Ceylon and the Straits Settlements. That was the start of Southeast Asia's natural rubber industry, and today most of the world's natural rubber supply comes from the Hevea tree.

As the Brazilian rubber monopoly continued to inflate prices, manufacturers in Europe and the United States began to invest in the small, but maturing, plantations in Southeast Asia that had come about because of Wickham's enterprise. Goodyear, which in 1916 had acquired 20,000 acres of plantation land in Sumatra, was among them. That year the company also established a sales office in Jakarta, then named Batavia, the capital of Indonesia.

The company had first expanded outside the United States in 1910 when it purchased a plant for tire manufacturing in Bowmanville, Ontario, from the Durham Rubber Co.; two years later it opened a branch office in London, England. So with its Eastern expansion in 1916, the young company had become truly international with operations in Canada, England, and Indonesia.

Goodyear's primary target from 1910 to 1920 was the U.S. automotive market. Management realized that to attain a healthy share of this market — hopefully the largest share — it needed a superior product, a tire that matched or surpassed all others. To develop this tire would require a corporate commitment to innovation, research, and development. That commitment was made.

An experimental department was established, later called the development department. It was staffed carefully with graduates from technical universities who got their initial on-the-job training from Litchfield, Seiberling, State, and others considered the company's original tire engineers.

New and improved products began coming from the development area in satisfactory volume. They included the S V solid tire for trucks that, unlike the common bolted-on truck tire, was pressed onto the rim; the All-Weather tread for auto tires; the Blue Streak motorcycle tire; an airplane tire; and miscellaneous industrial products. Nevertheless, development efforts continued to concentrate on auto tires, the main chance for real growth.

## Challenging Cotton Cord

Litchfield's new department had its first big challenge in 1912 when Goodyear decided to produce an alternative to the cotton cord tire fabric, a British invention brought to the United States by Goodrich in 1898.

In earlier pneumatic tires, the square-woven canvas fabric had little durability. The sawing action of cord against cord during flexing created friction that built up heat, shortening the life of the fabric and the tire. The heavier the load, the faster the failure.

The sudden success of the newly introduced longer-life cord tire prompted Goodyear to design its own cord fabric. The Diamond Rubber Co. of Akron, later acquired by Goodrich, had recently obtained U.S. rights to the Palmer cord tire, which had been developed in England, and put it on the market as the Silvertown Cord Tire.

It was built with two rubber-permeated plies of woven cords a little thicker than pencil lead separated by a cushion of rubber. All cords ran the same way with only a light cross thread that was broken during vulcanization. The Silvertown Cords were expensive — two to three times as much as other tires — but their construction almost eliminated the sawing motion and consequent heat buildup of a square-woven fabric and thus substantially increased tire life.

Paul Litchfield's heritage dated back to the sea for generations, and as a boy he had spent many hours in the shipyards of Bath, Maine, watching the boats come and go with the fishermen who manned them. One thing that stood out in his memories of those days was the resilience of the cord fishing nets and hemp lines used to moor the ocean liners.

He applied this phenomenon to the construction and use of the automobile tire and developed a revolutionary theory. A tire built with more plies of smaller cords would have greater flexibility, would be lighter, and would have enough strength to withstand rapid wear and abnormal impacts.

Goodyear quickly put Litchfield's idea to work. Using multiple plies at right angles to one another and smaller rivet cords, the company developed a radically new tire. Especially different was a boot — called the flipper — of tough cross-woven fabric that provided a stiffness where the bead attached tire to rim, the one place flexing was undesirable.

These were the first multi-ply, small-cord, straight-side, pneumatic tires for automobiles, which Goodyear called "No Rim-Cuts" to emphasize the protection at the base.

In the Seiberling tradition of backing conviction with action, Goodyear went all out to promote the new tire. Advertising, promotion, and publicity worked hand in hand. Ralph de Palma, a leading racing driver, used the tires in the 1913 Indianapolis 500 with success — and publicity. In 1914, all entries in the 500 were on Goodyear cord tires. The auto industry began to order Goodyear's sensational new product in volume, and by the start of World War I, it dominated the U.S. automobile tire market.

*Opposite, the first double-page ad for a U.S. company in a national magazine. The January 1, 1910, issue of the "Saturday Evening Post" carried this ad, representative of the imaginative, daring promotional techniques that helped double Goodyear's business each year from 1908 to 1912.*

United States Rubber, Goodrich, and Diamond shared tire industry leadership in early 1912. Goodyear, growing fast, was in fourth place. Late that year, Goodrich acquired Diamond, its Akron neighbor, and became the undisputed leader. U.S. Rubber, however, with 60 percent of its business in footwear, mechanical goods, clothing, and rubber sundries, was the largest rubber company.

## Early Advertising

In the years from 1908 through 1912, Goodyear had aggressively followed a policy of establishing itself firmly with the auto industry. It expanded dealerships into small communities neglected by competition, emphasized the quality of Goodyear tires, and dominated national tire advertising.

The company spent $110,893 on advertising in 1910 and $735,887 in 1912. So intent was Seiberling on keeping Goodyear's name and products in front of the public that he sometimes borrowed from future earnings to finance advertising and promotion.

The advertising push gathered full momentum in 1909 with full-page ads in such national weekly magazines as the *Saturday Evening Post, Collier's, Literary Digest,* and *Leslie's* and big ads in *Harper's* and other leading monthly magazines. Newspaper advertising was concentrated in cities with Goodyear sales branches, and selected top national trade journals were scheduled on a regular basis.

The company sent out a promotional package in a direct-mail campaign. It went to car dealers and owners and included a Goodyear ad reprint, the cover of the magazine in which it appeared, and a separate sales message.

Many industry skeptics belittled this method of promotion because the material reached only a small percentage of actual owners. Seiberling, Litchfield, and Stadelman, however, were counting on the wave of the future for the biggest share of an ever-increasing auto tire market.

Goodyear ads of that time were written by Claude C. Hopkins of the Lord & Thomas agency. Hopkins was considered the world's best-paid copywriter at $1,000 a week.

20 THE SATURDAY EVENING POST January 1, 1910 THE SATURDAY EVENING POST 21

**The Most Spectacular Landslide in Automobile History**

**36% of the estimated 150,000 new automobiles to be manufactured during 1910 will be equipped with GOODYEAR TIRES. The remaining 64% is divided among 22 established Tire Makers.**

**Bona Fide Contracts with the Prominent Automobile Manufacturers noted below call for the Amazing Total of 216,000 GOODYEAR TIRES. Tire contracts of such Magnitude have never before been heard of in the Automobile Business.**

**Our vast Dealers' and Consumers' Business will swell this Enormous Total by more than 50%. Those who would replace present tires with Long-lived, Trouble-proof Goodyear Equipment, should PLACE ORDERS EARLY to avoid Disappointment.**

Note this list of prominent automobile manufacturers who have contracted for Goodyear Tires for their 1910 output. Nearly all of the admittedly BEST cars in the world are on this list.

*American Locomotive Co.*
*American Motor Car Co.*
*Austin Automobile Co.*
*Babcock Electric Carriage Co.*
*Bartholomew Co.*
*B. C. K. Motor Car Co.*
*Buick Motor Co.*
*Cadillac Motor Car Co.*
*Cartercar Co.*
*Clark Motor Car Co.*
*Columbus Buggy Co.*
*Columbia Motor Car Co.*
*Corbin Motor Vehicle Corporation*
*Crawford Automobile Co.*
*Crow Motor Car Co.*
*Croxton-Keeton Motor Co.*
*Demotcar Co.*
*Elkhart Carriage & Harness Co.*
*Elmore Mfg. Co.*
*Fuller Buggy Co.*
*Great Western Automobile Co.*
*Indiana Motor & Mfg. Co.*
*Jackson Automobile Co.*
*Kissell Motor Car Co.*
*Marion Motor Car Co.*
*Maytag-Mason Motor Co.*
*Metz Co.*
*Mitchell Motor Car Co.*
*Oakland Motor Car Co.*
*Ohio Motor Car Co.*
*Olds Motor Works*
*Overland Automobile Co.*
*Packard Motor Car Co.*
*Peerless Motor Car Co.*
*Pierce Motor Co.*
*Pope Mfg. Co.*
*Simplex Motor Car Co.*
*Speedwell Motor Co.*
*Starr Motor Car Co.*
*Stevens-Duryea Co.*
*Waverley Co.*
*Welch Co. of Detroit*
*White Star Automobile Co.*
*York Motor Car Co.*

Not one of these firms could be influenced in placing a contract, by any other consideration than Supreme Quality. Any other policy would mean business suicide. For the day is here when the "Survival of the Fittest" looms up big in the Automobile Field.

Makers who have spent vast sums in building up a reputation for Quality and Performance in their cars must to a greater extent than ever before watch to their laurels and reject anything not the choicest and best of its kind which forms a part of the car they produce.

And the result is the largest landslide of orders ever placed with a single Tire Manufacturer in the history of the Automobile Business.

**Think What This Mammoth Quantity of Tire Orders Actually Means!**

Estimating our Dealers' and Consumers' business for 1910 as being only a third of our total business, the tires which we shall be required to furnish during 1910, if laid one on top of the other, would make a pile 28 miles high.

If they were laid tread to tread in a row, they would make a line of tires 228 miles long, or the distance between New York and Boston.

Or, cut across, and laid in tubes, end to end, they would make a pipe of rubber 682 miles long, which (following the shortest railroad route) would connect New York City with Toledo, Ohio, and a few miles to spare.

To make this tremendous number of tires will require 3,800,000 pounds of refined rubber and 1,300,000 yards of special Sea Island cotton fabric.

**Nothing Like it Was Ever Heard of Before in the Tire Making Business**

It confuses the mind — makes you almost dizzy to think of it.

And it makes us mighty proud — not only to have such a unanimous vindication of our claims that Goodyear Tires are and always have been the best obtainable because longest lived — nearest trouble proof and easiest to remove and replace should trouble come.

But also because of the triumph of Quality over Price, an unusual condition.

For it must be understood that Goodyear Tires are and always have been higher in price than other tires, and for a very good reason — they offer far more for the money.

**We feel that we are perfectly justified in becoming enthusiastic over this wonderful sales record of Goodyear Tires, made possible solely by the supreme excellence and merit of the Tires themselves.**

The reason why Goodyear Tires are so vastly superior to all others — and why they give such almost unbelievable mileage, often without punctures or trouble of any kind from one season's end to another, are given in brief on the opposite page.

If you are interested in making every penny invested in Tires count to the utmost, you will be wise to follow the lead of those who know, and use Goodyear Tires and no others.

**The Goodyear Tire & Rubber Co.,** Factory and Main Office: Seneca Street, **Akron, Ohio**

BRANCHES AND AGENCIES

**The Inside Reasons for the Supreme Excellence and Wonderful Popularity of GOODYEAR TIRES**

Goodyear Tires have always been specified by the Motorist who buys his Tires as painstakingly and intelligently as the successful merchant buys his stock of goods.

Because Goodyear Tires alone embody all the essentials which go to make up the theoretically perfect tire.

Many of the most vitally essential points are patented, which is a very good reason why Goodyear Tires have so long been in a class by themselves, unhampered by any real competition.

Convincing practical tests of unheard of severity, made under all sorts of conditions, have time after time convincingly proved every claim we have made for Goodyear Tires. When motorists become as keenly alive to their own interests in buying Tires — as manufacturers now are in selecting their equipment — or stop to think what our many strenuous tests of stock tires mean to them individually, it can result in but one thing —

**Goodyear Detachable Tires Will Dominate the Field**

The reasons for the supreme, unapproachable performance of Goodyear Tires — which are so perfect and trouble-proof that cars are often sold as second-hand with factory air still in the tires — are worth a careful reading by every motorist who has no money to waste.

We can touch but briefly on the more important features in this limited space. But remember this —

The Why and Wherefore of Every Claim of Perfection will be convincingly demonstrated to you by letter or in person whenever you say the word.

The foundation rock on which the Perfection of Goodyear Detachable Tires is built is in the design of the Tire itself. Refer to the cross section in the middle of the page. Note that the outer sides of the tire are straight, that no hooks are used at the base or feet of the tire.

This permits the use of a rim with a wide, rounding lip where it holds the tire in place. There is nothing there which could cause a rim-cut even though the tire were ridden absolutely flat, as is sometimes imperative.

Rim Cutting has cost the motorists of this country literally Thousands of Dollars.

**We Absolutely Guarantee Goodyear Detachable Tires Against Rim Cuts**

We control the patents on the Piano Wire Tape which prevents creeping, ripping out the valve stem and ruining the inner tube, unless kept continually inflated to the highest point.

In Goodyear Detachable Tires, tapes of piano wire are vulcanized into the base or feet on each side (see white spots in the section). These tapes contract with inflation. With but twenty pounds of air in the tube, the casing grips the rim so tightly that creeping or forcing off the rim is impossible. In all the years the Goodyear Detachable has been on the market it has never once come off the rim in use, when rim flanges were properly backed in place, even though ridden for long distances *absolutely flat.* When fully inflated, the casing exerts a pressure of 134 pounds per inch on the rim circumference, or a total pressure of 11,400 pounds for a 34x4 tire. There are 63 wires of piano steel in each base, each wire capable of standing a strain of 100 lbs.

**We Absolutely Guarantee Goodyear Detachable Tires Against Creeping or Coming Off the Rim in Use. No Tire Bolts Need be Used**

But freedom from Rim-Cutting is not the only advantageous feature of the straight-side Goodyear Detachable Tire — the only practical tire of its kind.

This construction makes it a far simpler, easier and cleaner task to remove or replace a tire in case of need than any other type of tire.

This construction also makes it possible for us to make our tires EVERY SIZE OVERSIZE — without the slightest danger of the tire cutting off the rim in use. Each Goodyear Detachable is 15 per cent larger than tires of other makes sold for the same size. So in straight side Goodyear Detachable Tires, you not only get Extra Quality, but Extra Size in every Tire you purchase.

This Goodyear "Straight Side" construction makes it far easier to remove a tire from the rims even when Quick Detachable Rims are used. First, because the beads or hooks on the feet of the ordinary tire stick or "freeze" in the channel for them in the rim. It takes prying and much pulling to remove such a tire. And second, because the base or feet of the beaded or hooked-foot tires must be wide, so that the tips or "toes" meet in the center of the rim. It is impossible without exerting much force to push the side wall inward, so that the Quick Detachable Rim may be unlocked and the tire removed. The tape of piano wire — not breadth of feet — holds the Goodyear Detachable firmly on the rim. Note the space between the feet in the cut in center column.

In the Goodyear Detachable the side walls press inward easily, and the tire itself slips off as easily as your cuffs and with little more effort.

So far we have spoken of the design of the Tire only. There are structural features of equal importance — dozens of them, each one big with meaning to the Tire user.

Taken in connection with the design of the Tire itself, they make Goodyear Tires so far in the lead of all others from every viewpoint as to be like a full-jeweled Howard, Waltham or Elgin when compared with the "Watch" which can be bought for a dollar bill.

As evidenced by the fact that when, NOW, for the first time in Automobile history, Supreme Quality to the minutest screw — (not Quantity of output) — is the price of continued existence in the Automobile field —

**The Majority of the Makers of the Best Cars in the World Contract for Goodyear Tires**

The premise is correct. Every Motorist will admit it. The conclusion is obvious. Why attempt to point the moral?

. . . .

You — who buy thoughtfully, carefully, intelligently, in other things —

Let us go further into this Tire Question with you.

If you are not already convinced that there is truly nothing "just as good," let us, in your own interest, and to save you money, give you facts and figures that will answer any question that may be hiding in the back of your mind, fully and completely.

Call at any branch house or write for our free book "How to Select an Auto Tire." Do this for your sake, not ours, and do it NOW.

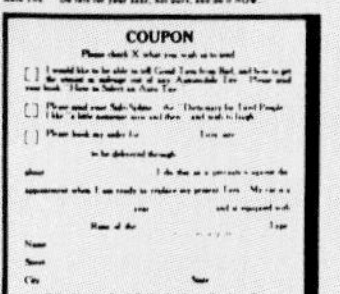

The Hopkins copy was loaded with "reasons why" Goodyear tires were superior and stressed advantages of the straight-side tire, which did not cut at the rim, in comparison with the clincher tire. Depending on an avid, growing public interest in cars and how they operated, he outlined the details of Goodyear's tire construction and its benefits to owners.

A full-page tire ad, the first by any tire manufacturer, had appeared on the inside back cover of the *Post* of February 6, 1901. Nearly nine years later, on January 1, 1910, Goodyear came up with another advertising first, the first double-page ad ever run by any U.S. company in a national magazine.

It was also in the *Post* and carried this headline: "The Most Spectacular Landslide in Automobile History." It listed the 44 automakers that had contracted to use Goodyear tires as original equipment, giving the company 36 percent of that market.

Goodyear's avalanche on the nation's magazines continued during the next few years with headlines like these.

Every Tire Oversize
An Amazing Record of Tire Perfection
The Original Time Tried and Reliable Straight Side Auto Tire
How to Avoid Tire Trouble
Cut Tire Expense

The Canadian subsidiary, acquired in 1910, had been advertising since its beginning. In the fall of that year, for example, it advertised that Goodyear tires were on more cars at the Canadian National Exposition than all other Canadian-made tires combined.

By 1912, Goodyear advertisements carried these headlines:

The Tire That Became The King
The Topmost Place in Tiredom
The Odometer — The Premium Salesman for Goodyear Tires
In 1906 a Speck on The Horizon. In 1912 the Biggest Factor in Tiredom

Goodyear alone then carried the straight-side banner as the bulk of tire makers remained loyal to the clincher. At that time, competition had gotten into the spirit of the sales arena and turned it into a free-swinging war of advertising. One company even carried an ad headed with this slur at the straight-side tire:

The Rotten Apple in the Middle of the Barrel.

The Goodyear spirit ran high in the years just before the United States entered World War I. Sales Manager Stadelman reflected this when he told his sales force that thanks to aggressive advertising, the name Goodyear on tires was as well-known to the public as Ivory on soap or Eastman on cameras.

The entire company was infused with confidence and drive, mainly because of its successes from 1908 through 1912 and the

dynamic leadership of a young and combative management. The Seiberlings were in their early 50s; Litchfield was 37, Stadelman 42. Many Goodyearites in the factories, offices, and field assumed heavy responsibilities while still in their 20s.

Promotions came swiftly as the company grew. Enthusiasm was high. Long blasts on the factory whistle greeted new production records. Sales branches sprang up throughout the nation. Salesmen became managers almost overnight.

Goodyear employees everywhere concentrated on getting the job done, forgetting the hours worked each day or on weekends.

Sales jumped from $1,794,822 in 1908 to $22,374,084 in 1912, profits from $100,933 to $3,001,294.

Investment in plant and equipment was six times greater in 1912 than in 1908.

*Early tire building techniques, circa 1920. Tires were built almost entirely by hand in the early years. Opposite, a tire builder applies the All-Weather Tread. Below, plies are stretched into position manually. Right, a young apprentice learns how a veteran repairs a tire tread.*

Issued capital in 1908 was $571,800 and bonds outstanding $170,400. In 1912, capital was $10,026,700 with no bonds.

Seiberling had no doubt taken some mammoth risks to finance new buildings and machines and to increase payrolls. But who knows how Goodyear might have fared if he had not?

Although Seiberling the entrepreneur and Litchfield the scientist were opposites in personality, they shared one characteristic of vital force in Goodyear's early progress. It was a positive, optimistic vision of the future of U.S. industry, a foresight that seemed to extend far beyond tomorrow and next year, upward even to the skies.

Litchfield's early confidence in aviation started the young and still-struggling company in aeronautics. In 1909, six years after Orville and Wilbur Wright had made their first flight at Kitty Hawk, North Carolina, Goodyear began building airplane tires. At the time, fewer than 100 planes operated in the United States; so the market was practically nonexistent. The future beckoned to Goodyear's management, though, and the company's Wing tire began to replace sled runners, skids, and bicycle tires on the flimsy flying machines.

During a 1910 trip to Europe with his factory superintendent Bill Stephens, Litchfield's interest in aviation as a future market was tremendously stimulated. The progress he saw in airplane building, plans for dirigibles, and other exciting evidence convinced him that the future of transportation was both in the air and on the ground. Two dirigible manufacturing companies, in fact, had been organized in France and one in Germany, and both a plane and an airship had flown across the English Channel.

On learning that the North British Rubber Co. at Edinburgh, Scotland, had developed a successful process for spreading rubber over fabric, Litchfield and Stephens made plans to visit its plant. During meetings there, the two sides exchanged specifications: the straight-side tire for the company's new process.

The Goodyear envoys also returned home with two Scots to operate the spreading equipment. Both men — Fergie Ferguson and Andy Aikman — were Goodyear employees for the next 38 years. They ran the spreader department as their own domain, and management, confident of their dedication and expertise, seldom bothered them.

Late in 1910, the Wright brothers adopted the new Goodyear fabric, which quickly gained in popularity and was soon on most airplanes flying in the United States. In 1913, it was specified as standard equipment for U.S. Army and Navy planes and was used on military aircraft for several years before replacement by metal wings and a fabric treated with dope.

Goodyear's first association with lighter-than-air craft was in 1910 when Walter E. Wellman, a one-time Ohio newspaper publisher and later Washington correspondent for the *Washington Record-Herald* newspaper, visited Akron in search of financial backing for a dirigible flight to cross the Atlantic. Earlier that year, his dirigible *America* had crashed in the ocean on a first attempt. Despite being blown off its course, *America* had covered 1,008 miles, a record for air travel. Wellman was confident that a bigger dirigible could succeed in an Atlantic crossing.

Although the recovered gondola of *America* hung in Goodyear's Akron garage for more than a year, no record exists that Goodyear backed the new dirigible, named *Akron*. The *Akron*'s rubberized fabric was manufactured and cut to shape in the Goodyear factory, though, and sent to Atlantic City for assembly. The big ship was 258 feet long and 47 feet in diameter and held 400,000 cubic feet of hydrogen.

The transatlantic flight of *Akron* began in Atlantic City at 6:15 on the morning of July 2, 1912. It ended exactly 23 minutes later when the ship exploded and fell to the ocean in fragments. The crew of five was lost and the cause of the tragedy never determined.

Seiberling and Litchfield nevertheless maintained confidence in the future of travel by air.

Two years earlier, Litchfield had hired a pair of young engineers to staff the aviation section of his development department: Ralph Upson, recently graduated from Stephens Tech, and R.A.D. Preston, from MIT. The two soon were flying, designing, and helping build balloons. In 1913, they entered and won the James Gordon Bennett International Balloon Race from Tuilleries Gardens in the heart of Paris to England. With the eruption of war in Europe, lighter-than-air competition on the continent and in Great Britain was halted, and military forces began using aircraft in warfare.

## Seiberling's Housing Plan

Goodyear employed about 6,800 people in 1912. Most of the factory and salaried employees in Akron lived on the east side of town, but many of them and others who lived farther away had to travel a half hour or more to get to work. F. A. Seiberling developed a housing plan to reduce the commuting time of employees and to give them "an opportunity to become home owners at as low a cost and by as easy means as possible." He announced the plan at a meeting of factory foremen in July 1912.

In his own name, he had acquired a large tract of land on hills east of the plant to develop into a model residential area for Goodyear personnel. A 10-minute walk from the headquarter-factory complex, the strictly residential property would be subdivided to take full advantage of the rolling terrain. The area would have a complete system of public utilities, plus sidewalks and paved streets, and

buyers could choose from 20 different architectural plans for three general kinds of houses: single unit, double unit, and bungalow.

The development, called Goodyear Heights, would be started with 100 houses. Down payments were not required, and monthly payments were but slightly higher than rentals for some of Akron's low-cost housing. Financing came from 20-year first mortgages held by Metropolitan Life Insurance Co. and 12-year second mortgages held by Goodyear.

To prevent speculation, 25 percent was added to cost, and payments were based on that increased price for the first five years. If the owners were still Goodyear employees after that time, the amount initially added, with interest, was deducted from the mortgage so that the employee got the house at cost plus interest.

By May 1914, 90 dwellings had been completed and were occupied. By September 1916, 153 were finished and all 400 lots in the first tract sold. Most properties in that tract had been purchased for about $3,500 each.

In 1913, Litchfield established a training plan — called the Flying Squadron — that became the nucleus of the company's personnel development program. This plan was continued for a half century, except during the difficult times of 1920 and 1921. Its initial aim was to give broad experience to 50 top factory workers so they would be knowledgeable about every manufacturing operation. When trained, they could enter a lagging department to speed it up or an overburdened area during high-volume production.

*Opposite, the Akron-Boston Express of 1917 dramatically demonstrated the durability of Goodyear's new pneumatic truck tires. Top, the first Flying Squadron, 1915. Bottom, the Goodyear Heights bus line transported workers home to F. A. Seiberling's model community.*

According to Litchfield, the Flying Squadron provided "an opportunity for a broad education in the different operations of a rubber factory; and those who demonstrate their ability will be in line for higher positions."

Although the squadron was initiated mainly to ensure balanced production, its greatest benefits were in personnel development. In the prewar years and for the next 50, a large percentage of squadron graduates rose to top levels of management. The early three-year courses concentrated on production and engineering, but also included such subjects as effective speaking, management, mechanical drawing, mathematics, and physics.

Starting in 1920, squadron members could supplement on-the-job training with courses at the Goodyear Industrial University, which had been established in Goodyear Hall. A six-story structure started on East Market in 1917, the hall was not completed until 1920 because of the war. Litchfield conceived the idea for this facility, a place for employee education, sports activities, and pastimes.

Most nonsquadron classes focused on business-related subjects, such as bookkeeping, typing, shorthand, public speaking, credit, and personnel relations. The curriculum was soon extended to more general areas, offering courses equivalent to some in the local high schools. More than 20,000 Goodyearites attended classes at the corporate university in its first two years.

The Litchfield awards program, established just before World War I, was a useful adjunct to the overall training program. Called the Paul W. and Florence B. Litchfield Awards, it recognized exceptional performance in the production and engineering squadrons. Expanded gradually, the program ultimately provided annual awards for certain outstanding achievements: best company suggestion, top-ranking member of the Industrial University, and unusual service beyond the call of duty. Awards also were given to leading Boy Scouts in troops sponsored by Goodyear.

In the postwar period, the scope of the awards program broadened to include the best domestic and export salesmen, the best retail store manager, top-ranking apprentices, and additional awards to outstanding members of the Boy Scouts of America, namely Sea Scouts and Air Scouts. Winners' names were inscribed on bronze panels in the lobby of Goodyear Hall.

Amended from time to time to meet changing conditions, the Litchfield awards program continues as a corporate project.

## A Factory Hero

The identification of high-potential personnel generally was the responsibility of top managers who based their evaluations on observations of employees at work. Attitude was considered an important factor. Most early judgments were informal and subjective, but they seem to have worked. The case of Clifton Slusser is a good example.

While secretary to Bill Stephens, Slusser performed outstandingly in the cleanup following a disastrous flood in March 1913. The usually placid Little Cuyahoga became a raging torrent as a result of heavy spring thaws and intensive rains. Overflow water flooded the plant, in some places rising eight feet above floor level.

Slusser took charge of some hastily organized cleanup crews and restored his areas to normal well ahead of anyone else. Further, he had his workers scour the riverbanks for miles downstream where they collected many bales of rubber carried away by the floodwaters. His thoroughness and organizational abilities made a deep impression on Litchfield, who immediately marked him for future management responsibility. Slusser eventually rose to vice-president in charge of production and because of his move up through the ranks became a hero to the production force.

From 1912 to 1916, Goodyear continued to increase its share of the auto tire market. Production, which had reached 35,000 tires in 1908 and exceeded a million in 1912, rose to 1.5 million in 1913. Goodyear's greatest strength then was in original equipment supplied to the major auto manufacturers. A tally of the 5,916 cars exhibited at the big motor shows of 1913 revealed that 39.08 percent were on Goodyear tires, 26.52 percent on U.S. Rubber, 17.59 percent on Goodrich and Diamond combined, and 6.66 percent on Firestone.

Although the founding of Firestone in Akron by farm-born Harvey Firestone had attracted little attention in 1900, the company was definitely a factor in the industry by 1906. Some of its early impetus came from tire sales to Ford that year, and for the next seven decades Firestone would be Ford's number one tire supplier. The meeting of Henry Ford and Harvey Firestone during a business transaction in 1905 developed into a lifelong friendship.

U.S. Rubber, Goodrich (which had merged with Diamond in 1912), Goodyear, and Firestone had emerged as the Big Four of the tire industry by 1913. Two years later, William O'Neil, son of the department store owner, and Winifred E. Fouse, who had been credit manager of Firestone, established the General Tire and Rubber Co., which by 1935 was the nation's fifth largest rubber company.

*Opposite, top, Goodyear's first delivery truck transported tires in large wooden barrels to Akron's train station, returning with raw materials for production. Far left, Plant I's engine room during the flood of the Little Cuyahoga, 1913. Left, office workers outside Goodyear's factory office, 1917.*

In 1916, Goodyear became the world's leader in tire sales, and it made the most of this leadership in advertising and promotion. Based on its number one position and public preference, the company adopted a new slogan: "More People Ride on Goodyear Tires Than On Any Other Kind." This statement has been a Goodyear claim ever since.

## An Enduring Message

Another corporate slogan, "Protect Our Good Name," had been introduced in the preceding year. Directed at the growing Goodyear family, it too has remained with the company.

"Protect Our Good Name" was coined by Theodore F. McManus, chief copywriter for Erwin-Wasey, which had succeeded Lord & Thomas as Goodyear's advertising agency.

McManus wrote an ad pointing out that a company was judged by what it did, how it performed, the quality of its products, and its character. It emphasized that the company's future and the futures of its employees are the responsibility of the employees themselves. The company's good name, thus, was in their hands.

Seiberling and Litchfield were both impressed with this ad — especially its primary slogan — and had it printed as a full-page promotion in the *Post* on November 6, 1915, and in newspapers around the country. Signs bearing the phrase "Protect Our Good Name" were posted throughout the factory and in all branch offices. Nearly 70 years later, this slogan — translated into many languages — occupies prominent positions on Goodyear factory and office walls throughout the world.

McManus was succeeded at Erwin-Wasey by Arthur Kudner, who would create some of Goodyear's most effective ads. One of his first, published in the *Post* on February 10, 1917, was designed to combat the "mileage guarantees" offered by many other manufacturers. Most tire companies guaranteed 3,500 miles to a tire. The ad's headline was: THERE ARE NO MILES IN A BOTTLE OF INK. The 21 paragraphs of copy included some unusual advertising messages for the time.

> One of the pleasantest fictions of the tire business is that the manufacturer can repair with a pen what he has failed to accomplish in his factory.
>
> But the difference between mileage as adjusted over a counter and the mileage as delivered on the road is pressing hard for recognition upon the intelligence of the American automobile owner.
>
> And the guarantee fast is coming to be seen for what it is — a confusing and unnecessary formality in the case of a good tire, a disappointment and a delusion in the case of a bad one.
>
> There is no definite mileage guarantee behind Goodyear Cord tires.
>
> As for that, there is no definite guarantee behind a gold piece.
>
> None is needed in either case, for both are recognized as measures of value. Both embody a positive dimension of worth.

Kudner left Erwin-Wasey in 1935 to start his own agency and took the Goodyear account with him. His firm handled the company's tire advertising for nearly 40 years.

Goodyear's good industrial relations — which in those times might better be described

*Left, an advertisement from the March 1916 "National Geographic" magazine. Opposite, right, a worker enjoys a 25¢ lunch in the new Goodyear cafeteria, 1913. Opposite, left, Goodyear's factory hospital, established in 1912, had six beds and a nurse who was present during day and night shifts.*

as human relations — were taking shape in the century's second decade. The payroll grew from 2,500 in 1910 to 6,000 in 1913, and close personal relations between top management and most of the employees were no longer possible. Litchfield, who was in charge of personnel, established a labor department to develop and maintain good employee relations.

## Enlightened Human Relations

The earliest big strides in industrial relations were taken in 1912. The first issue of an employee publication, the *Wingfoot Clan,* was published on June 1 that year, and Litchfield wrote the editorial:

> We wish to make conditions in this factory as near right as possible, to make it a desirable place to work, to make each employee feel that he can get as high or higher reward for his services here as they will bring elsewhere, and we wish the cooperation of each employee to help carry out this program.

A factory hospital was established under Dr. J. S. Millard. It was reportedly only the second industrial hospital in the United States, preceded by one at International Harvester. A factory canteen began daily operations in 1912, and an athletic field named for F. A. Seiberling was provided for employee sports activities.

In 1914, Goodyear presented its first gold service pins in recognition of employment for 5, 10, and 15 years. The first 15-year pins went to Edward Hippensteal, George Swartz, Albert Cunnington, and Farran Hills, all of the factory force, and Clara Bingham and the Seiberling brothers of the general office.

Litchfield's enlightened attitude in industrial relations then was manifest in many ways. He had a hand in the formation of the Goodyear Relief Association in 1909 — a fund created by employee contributions and made available to sick or injured Goodyearites — and in 1915 he pledged personal funds to supplement the company's welfare activities. On July 15, 1915, at a dinner honoring his 15 years of service, Litchfield made a surprising disclosure.

> Now, this being my 15th anniversary in charge of the factory force, I would like this evening to share my appreciation of the support and help which the factory organization has given me.
>
> This evening my wife joins me in making a gift to the factory organization. We wish to give to the factory, in the form of a fund, all the salary which I have drawn in the past 15 years, with interest added to bring the amount to $100,000.
>
> This is not to be used in cases of sickness or accident, as our Goodyear Relief Association fully takes care of that, and every employee should belong to the association. It is not to take care of their old age, as I hope some day in the future a pension fund will take care of that.
>
> It is to promote loyalty, teamwork, and efficiency in the organization, to give you business responsibility in the handling of capital. Instead of stating what shall be done with it myself, I give it to you men as trustees, with entire confidence that you will use it both wisely and justly.

*An active interest in sports was an important adjunct to working on the job for many early Goodyearites. Goodyear Hall's huge gymnasium accommodated the major "varsity" sport, basketball, almost from the hall's opening in 1920. A high percentage of employees participated in the company's recreation program, which had spread to all factories by World War I and continued to grow through the years.*

*Left, a 1920 basketball enthusiast. Above, Goodyear Heights girls' baseball team, 1915. Opposite, the fabric tire department bowling team of 1918-1922.*

The Service Pin Association was formed to manage the fund, which, with the consent of the directors, was used to buy 1,000 shares of Goodyear common stock at par.

The fund was managed wisely. Sixty-five years later — without any infusion of additional money — it was providing a dozen scholarships (in total) at the University of Akron and Kent State University for children of employees or retirees. The scholarship program was started at Akron U. in 1963 and at Kent State in 1970.

The Service Pin Association also administered the Litchfield awards program, which recognized outstanding employee achievement above and beyond regular duty. Although the association had nothing to do with service pins, its name went unchanged until 1982 when it became the Paul W. Litchfield Awards Association.

As the company grew and its operations became more complex, management was given a more definite structure than provided in the original organization. At the stockholder's meeting of December 16, 1915, the Code of Regulations was amended to provide that "The standing officers shall be a President; three Vice Presidents, all of whom shall be Directors; a Secretary; Assistant Secretary; Treasurer; Assistant Treasurer; Second General Manager."

Elected to fill these posts were : F. A. Seiberling, president and general manager; C. W. Seiberling, vice-president; G. M. Stadelman, vice-president and sales manager; P. W. Litchfield, vice-president and factory manager; A. F. Osterlich, secretary and assistant sales manager; W. E. Palmer, assistant secretary and assistant treasurer; F. H. Adams, treasurer; and H. G. Blacburn, second assistant treasurer.

Goodyear had been manufacturing nontire products in small volume since its beginning, but in 1913 management decided to diversify and expand these operations. Late that year, a complete mechanical goods department was set up and, using newly purchased equipment, began to produce rubber belts, hose, packing, mats, and other molded goods.

At the outset, Goodyear lacked expertise in the large-scale production and marketing of these goods, which later became known as

industrial products. Any shortcomings the company had, though, were overcome quickly by the addition of three experts in those areas — I. R. "Bill" Bailey, "Pop" Metzler, and Hal Campbell — all from Goodrich. Bailey and Metzler, both early employees of Diamond Rubber, had gone with Goodrich in 1912 when the two companies merged. It was said that in their first year with Goodyear, Diamond product samples were used to sell certain Goodyear mechanical goods.

Some insiders claimed the three veterans were enticed from Goodrich by promises of key positions in Goodyear's newly enlarged department. Others said Bailey, who was the assistant sales manager for mechanical goods at Goodrich, came to Frank Seiberling and sold him on going into the mechanical goods business in a big way and in hiring the three Goodrich men.

Whichever story was true, raiding other companies for competent specialists was common in those days. Metzler and Campbell led Goodyear's development operations, and Bailey was made a sales manager.

According to Herman Morse, who joined the company as a mechanical goods development engineer and headed the development operations from 1920 to 1961, Seiberling moved fast into mechanical goods despite Litchfield's protests. Nevertheless, Litchfield supported the move once it was made, although mechanical goods never got as much of his attention or interest as tires.

Morse became a Goodyearite in 1915 as did Walter E. Shively who was assigned to tire design and development. For more than four decades, they worked to improve Goodyear products, Shively in tires and Morse in industrial products. For 25 years after retirement, both regularly attended Goodyear functions in Akron.

Early mechanical goods development at Goodyear was more a matter of pragmatics than engineering theory. Metzler, who for a while ran both the development and production teams, had been taught largely by experience and always carried a little black book that contained all compounds being used at the time. He and others with similar backgrounds were called drugstore compounders. The first college-trained development man in the department was Karl Kilborn, a 1911 graduate of MIT. Morse, who succeeded him, was also an MIT alumnus.

The mechanical goods sales group had no revolutionary products to offer and was forced to build its business by chewing away at competition. The going was difficult, and the department lost $500,000 in its first two years. But it made steady progress thereafter.

Goodyear's first attempt to gain control over raw materials was in 1913 when for $100,000 it purchased a cotton mill that had been shut down for two years, at Killingly, Connecticut. It began operations there late in 1913 and was running the mill at capacity by the spring of 1914, producing one million yards of square woven fabric per year.

Killingly was a factor in the improvement of Goodyear's square-woven tire fabric and in the development of cord fabric. It also provided a laboratory for Litchfield and the experimental department in developing multi-ply cord for automobiles and pneumatic tires for trucks.

The competitive urge and aggressiveness of Goodyear's hard-driving managers gave the company a head start in the use of a manufacturing machine that revolutionized early rubber manufacturing and has remained an essential part of rubber goods production. It was the Banbury mixer.

Until 1916, the mixing of rubber with other compounds to make materials suitable for conversion into tires and other rubber products was laborious and time consuming. Chunks of rubber were fed into two steel mills rolling in opposite directions and at different speeds, a process that broke down the rubber and compressed it into a long sheet before other compounds were added. The environment was hot and dirty, the method slow; to mix a batch of rubber and press it into sheets required 25 minutes.

## Birth of the Banbury

F. H. Banbury, a young engineer with the Birmingham Iron Foundry in Connecticut, had been working on the development of an automatic machine into which rubber and compounds could be dumped, then mixed by rotating blades. This machine would speed up the mixing process and occupy much less space than the steel mills.

Bill State, master mechanic at Goodyear, persuaded Banbury to ship one to Akron for testing under Banbury's personal supervision. An agreement was made, and the first of these machines arrived in Akron for several months of hard testing.

Banbury stayed with his equipment throughout the test period, working day and night and often sleeping on a cot in the factory's first-aid room. In the end, his machine proved to be a tremendous time and motion saver, requiring less than a tenth of the time formerly needed for the mixing process. State ordered 10 more as soon as the first tests were conclusive, and Goodyear enjoyed a lead of about two months in the use of this new marvel.

As the nation's drive into the Automotive Age accelerated, the production of trucks and buses increased rapidly — from about 25,000 in 1913 to more than 92,000 in 1916. Most rode on solid rubber tires, although some light trucks used pneumatics.

In 1913, Goodyear introduced the S V truck tire. It was a solid rubber tire pressed directly onto the wheel and sold at a 10 percent premium. The hydraulic press equipment used in installation was expensive; so the presses were installed in company-owned branches until a chain of truck-tire dealers was gradually established.

Heavily promoted and with a guarantee that the S V would give lower cost per mile than competing tires, the new truck tire — the last of Goodyear's solids — was a quick success. But the pneumatic tire for trucks gained popularity far more quickly.

Litchfield's engineers had discovered that the tough cord fabric that had revolutionized passenger tires was equally suitable for trucks. Additional plies were required for greater strength, but they did not lessen the tire's resilience. At a sales conference in October 1916, Litchfield announced the new cord pneumatic tire and said it would be demonstrated on trucks that would haul company products, mostly shoe soles and heels, to East Coast branch offices and return with loads of fabric.

On April 9, 1917, a five-ton Packard truck outfitted with pneumatic tires front and rear set out from Akron for Boston. Roads were poor, weather was bad, and the journey required 19 days. The return trip was made in five. The Akron-Boston shuttle was then run regularly and given the name "Wingfoot Express." It was a forerunner of the interstate trucking industry. More trucks were added, and the fleet comprised two Packards and two Whites. The round trip was made regularly in five days, a schedule the railroads could not match.

In May 1918, Goodyear put on a spectacular demonstration of the possibilities of long-distance hauling by trucks when five 5-ton vehicles carted 18 tons of tires from Akron to Chicago, reloaded with 18 tons of Red Cross hospital supplies, and transported them to dockside at Baltimore. The Akron-Chicago trip took 35½ hours, the Chicago-Baltimore leg 100. From Baltimore the fleet drove to Trenton, New Jersey, to pick up a cargo of wire for Akron. The highly publicized haulage feat, the first of its kind, was completed in one week.

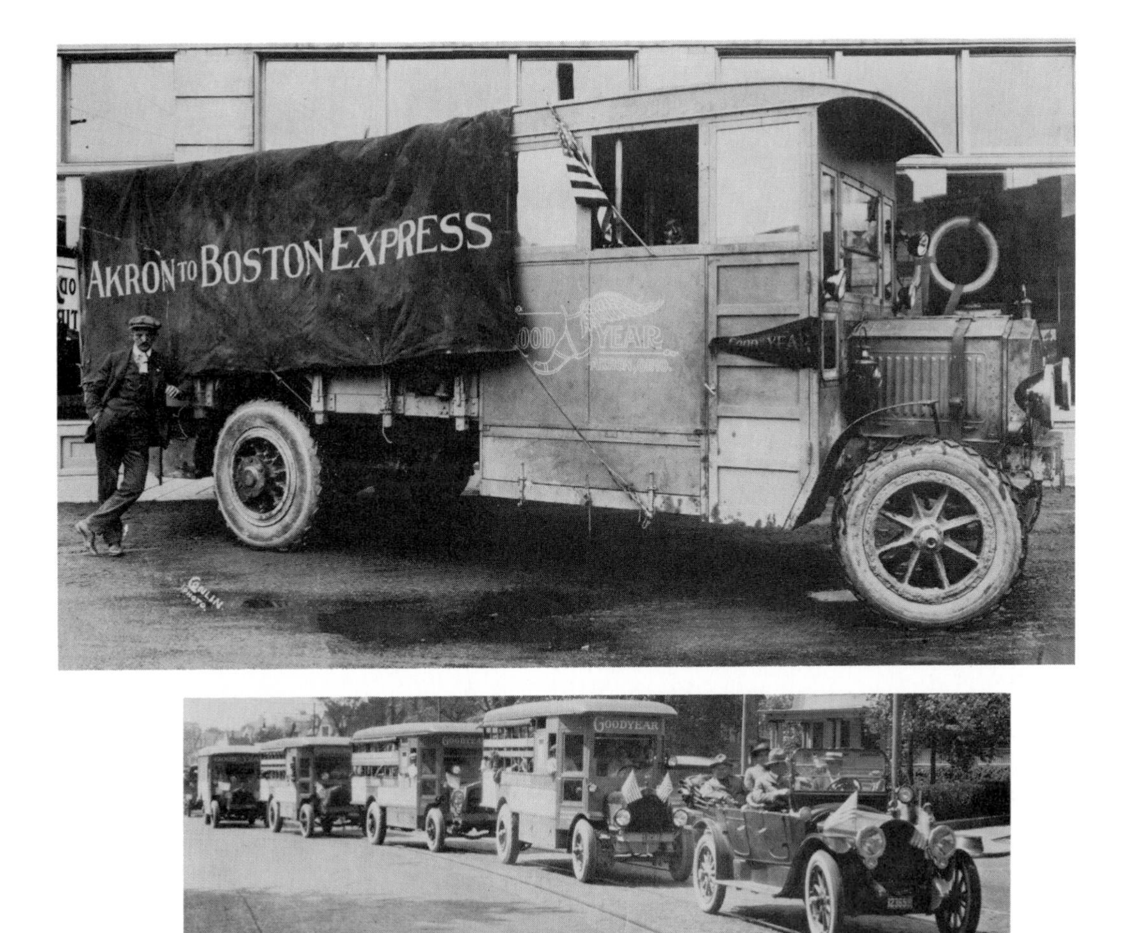


*Marketing joined with motoring when Goodyear took to the roads to promote its pneumatic truck tires. Opposite and above, bottom, 50 Boy Scouts made the trip when the "first truck train in America" traveled from Akron to Philadelphia, New York, Boston, and Washington in July 1918. Top, the Akron to Boston Express, forerunner of today's interstate trucking industry, regularly made its round trip in only five days!*

To demonstrate the troop movement potential of pneumatic-tired trucks, the company sponsored what the *Wingfoot Clan* described as the "first truck train in America." It was a July convoy of five trucks carrying 50 Boy Scouts on a two-week journey from Akron to Philadelphia, New York, Boston, Washington, and other cities before returning home. The young passengers were selected from Akron's 1,500 scouts by a competition in crafts.

Three trucks transported the boys, and the other two carried camping equipment and a field kitchen. The nearly 3,000 miles were covered in nine days, and the scouts enjoyed a five-day vacation in Massachusetts. The entire trip was made without a single blowout, a fact that Goodyear publicists emphasized to the news media.

The convoy was greeted at Boston's statehouse by the lieutenant governor of Massachusetts, Calvin Coolidge. In Washington, President Woodrow Wilson and Paul Litchfield met the scouts at the White House.

On their return to Akron, the young travelers, who had slept in the trucks and cooked out along the way, were greeted by thousands of well-wishers on Market Street just outside the Goodyear complex. The Goodyearite in charge of the convoy, Edwin J. Thomas, rode and slept in the lead truck. He had joined the company two years earlier, was assistant to Litchfield's secretary, and was hardly older than some of the scouts. In 1940, he would become the company's eighth president.

# Transcontinental Trucking

Later that year, in a high-visibility stunt that set the tone of Goodyear promotions for many years, the company demonstrated the feasibility of transcontinental trucking. This was the first organized attempt to cross the United States by truck.

On September 1, 1918, two big pneumatic-tired trucks, accompanied by a pilot automobile, left Boston loaded with fabric. They delivered their loads in Akron, reloaded with airplane tires for West Coast flying fields, and arrived in San Francisco on September 22.

*Roads and bridges — not the trucks or tires — were the biggest obstacles to a speedy trip for the "Transcontinental Motor Express" of 1918. While crossing Wyoming, the trucks reportedly broke through 52 crude bridges that had to be propped up before the express could move on.*

Each truck had two men who alternated driving and sleeping so that the convoy could run day and night. Including the return trip via Los Angeles, the convoy traveled 7,763 miles; 71.5 percent of the route was unpaved.

World War I hastened the development of pneumatic tires for trucks. Most Army trucks were on solid tires, but many used pneumatics. When heavy snows slowed several railroads in the winter of 1917-18 and solid-tired trucks could not cross some of the eastern mountains, the federal government made use of the Goodyear trucks to transport urgently needed war materials from the Midwest to Atlantic Coast seaports.

Following the war, American tire companies continued to concentrate on the manufacture of solid truck tires, but Goodyear moved to pneumatics on a large scale. Litchfield was unconcerned about dropping to third in solid-tire production; he was convinced they were on the way out. By 1926, pneumatics outnumbered solids in U.S. production. By 1930, they outnumbered solids 10 to 1.

In 1919, Goodyear set up a highway transportation division to study over-the-road trucking opportunities and problems. One finding was that as trucks became larger, they also became higher, more difficult to load, and tougher on tires. So in 1920, the division developed and constructed a six-wheel vehicle with chain-driven tandem wheels. It provided increased payloads, greater road speed, more efficient loading and unloading, and was easier on tires than the four-wheeled trucks.

The division manufactured several more of the six-wheelers, which went into the Wingfoot Express operation, and a six-wheel bus for use between the factory and Goodyear Heights. By the early 1920s, the practicality of long-distance hauling on six-wheel pneumatic-tired trucks had been effectively demonstrated. Its mission accomplished, the Wingfoot Express was retired.

Aware of the great potential in pneumatic tires, Goodyear's management began a search for a reliable supply of long staple cotton, an important component of the new tires. The two main sources of a high-grade supply, Egypt and the Sea Islands off the Georgia-Carolina coast, appeared insufficient to meet expected needs. Add to that the threat of war cutting off supplies from Egypt and the pesky boll weevil plaguing the Sea Islands crop.

## Cotton: The Desert Blooms

Litchfield looked to Arizona where U.S. Department of Agriculture experiments had produced long staple cotton of excellent quality. He went to Phoenix in 1916 and bought 22,000 acres in two tracts in the Salt River Valley. One, called Litchfield Ranch, encompassed 14,000 acres southwest of Phoenix. The other, named Goodyear Farms, took in 8,000 acres southeast of the city. A third ranch, 12 miles northeast of Litchfield Ranch, was purchased in 1921, bringing Goodyear's total holdings in the Phoenix area to 36,000 acres.

The land was cleared, hundreds of miles of canals and ditches were built to convey the ample supply of underground water, and the cotton was planted. When harvested, it proved to be of a high quality, and the company was assured of enough cotton to build pneumatic tires for many years to come. In the prewar years, Goodyear advertised its Arizona cotton — named Sarival after Salt River Valley — as a positive element in its pneumatic tires.

Developing the cotton ranch, especially in the haste that Goodyear demanded, was a large-scale construction job. Litchfield described it this way in *Industrial Voyage,* his memoirs published in 1954.

> It was nothing to worry the contractors of that era, who were building railroads across the country through mountains and over rivers. Give them enough money and equipment and tell them you want the job done, and they will do it. It will cost you more than if you are willing to go more slowly, but in this case we could not wait. Time literally meant money.
>
> Almost overnight an army of men started moving in, hundreds of Mexicans who threw up brushwood and adobe shelters, and a still larger force of adventurers drawn from all over the West by the lure of good pay, men who went from one big construction job to the next. We soon had 2,000 men on the job — and 1,300 mules at the peak, more than half the mule population of the state.
>
> Clearing and grading methods were primitive, but effective. Big Caterpillar tractors rumbled in from the Coast, were hooked up in tandem with a length of railroad steel between them, and set off across the desert. Other gangs were drilling wells, putting in pumps, building power lines, laying out a network of concrete canals and irrigation ditches, building highways, town sites, even a railroad.
>
> The property was so large and the roads so bad that we had to buy an airplane to get over it.
>
> I might say that we harvested cotton the next fall from 4,000 acres of land that formerly had been desert — cotton that grew taller than a man's head and was comparable in quality to the best Egyptian grades.

## Birth of a Famous Resort

As construction progressed, a headquarters town named Litchfield Park was laid out about 18 miles from downtown Phoenix. It included broad streets lined by newly planted palm, citrus, and pepper trees, and buildings were fashioned in the Hopi Indian style of architecture. A few houses were grouped together for the ranch management staff, and a small company hotel called the Wigwam was built to accommodate occasional visitors.

In a few years, the Wigwam had gained a statewide reputation for its hospitality and gardenlike surroundings. So the company decided in 1924 to expand it into a winter resort with golf courses, swimming pool, stables, and other amenities. Its reputation and size continued to grow; today it spreads over 900 acres and includes riding trails, three 18-hole golf courses, a tennis complex, and 75 acres of living accommodations and grounds. The Wigwam is one of a very few resorts in the United States to have the top rating of five stars in the *Mobil Travel Guide.*

When the United States entered World War I on April 6, 1917, Goodyear was second only to U.S. Rubber in total volume and first in the manufacture and sale of tires. Industry leaders and sales volumes were: U.S. Rubber, $171,135,222; Goodyear, $103,558,669; Goodrich, $87,155,072; and Firestone, $61,587,219.

The strongest competition for the Big Four came from Fisk, Ajax, Lee, Kelly-Springfield, and Miller. These companies sold some tires to the auto manufacturers, but, like most of the industry's smaller members, they concentrated on the replacement market — sales to the public.

*The 146th Regiment parades down Akron's Market Street in April 1919 to celebrate the end of World War I and the Allied victory. Flags draped from Goodyear's main office buildings and applauding crowds salute the returning doughboys.*

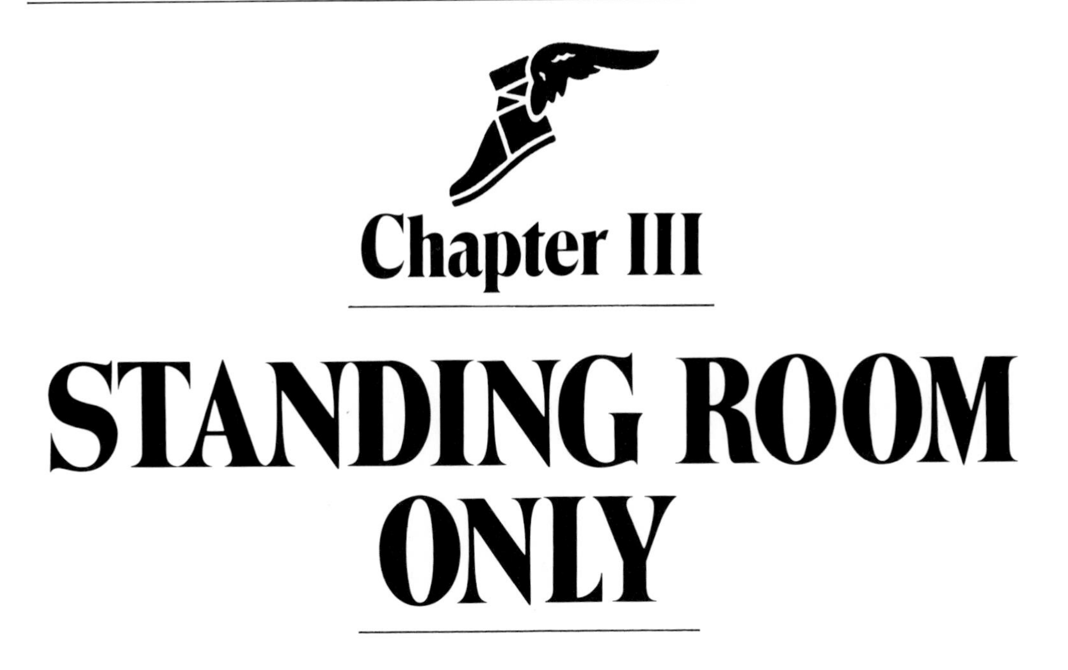

# Chapter III

# STANDING ROOM ONLY

The First World War, which started a month after the assassination of the Grand Duke and Duchess of Austria in Sarajevo, Bosnia, on June 28, 1914, brought on great changes in the U.S. economy and its new automotive and rubber industries.

Spurred by demands for supplies to Europe, factories moved into high gear. Employment soared and so did wages. The mass-produced automobile was no longer just for the wealthy and privileged, because almost anyone with a decent job could now afford a car.

Auto production rose from 543,679 in 1914 to 1,756,790 in 1917, and a new prosperity came to the tire companies. As this business boomed, so did the Rubber Capital of the World. Akron's population more than tripled, jumping from 69,067 in 1910 to 255,000 in 1929.

Thousands migrated there from other parts of Ohio and from the farms and mountain areas of West Virginia. "Akron, Capital of West Virginia" was a standard witticism of the time.

Seiberling, Litchfield, and their associates made the most of the boom opportunities. Expansion was a key word for Goodyear in the war years. Total investment in real estate, buildings, machinery, and fixtures in 1916 was $12,689,055. By 1920, the investment for the Akron plant alone was $60,307,035.

The company entered the war program in earnest, and it was more than ready for the call. Goodyear manufactured the best solid tire on the market; its pneumatic tires for trucks were proving to be an excellent product; it was the only company that designed tires specifically for airplanes; and it was one of few manufacturers with lighter-than-air technology.

Seiberling pointed out, "The automobile fits in for the direct purpose of the war to an extent we have not yet discovered. More cars will be built than we ever dreamed of."

Concerning aeronautics, he said: "We seem to be the only organization in the United States that could quickly take care of certain needs of the government. We are going to go on and do for our government all that we can in the direction of giving it its needed balloon equipment for the purposes of the Army and the Navy."

Production was built up quickly to supply these needs. The main products were automobile tires, solid and pneumatic truck tires, balloons, gas masks, mechanical rubber goods, and rubber parts for airplanes. Production of solid tires for trucks zoomed from 250 a day in 1916 to nearly 4,000 a day by war's end.

Goodyear made 1,000 balloons of various kinds and 60 airships during the war. Gas mask production totaled 715,000, along with 4,750,000 parts, mostly valves.

Aeronautical sales in the 1917-18 fiscal year totaled $4.8 million. The small department that built tires for airplanes mushroomed, producing a large percentage of tires used on the 11,000 military planes built during the war. As the planes grew, so did the tire sizes; one-ton bombers of the war's early days were gradually replaced by bigger and bigger planes, topped by 15-tonners at the signing of the Armistice in November 1918.

Goodyear's most spectacular products in the war years were the balloons used for observation and artillery reconnaissance and the blimps used in antisubmarine operations. The slow-flying blimps could move low over water and were ideal for searching out enemy submarines along coastal areas.

The first Goodyear blimps were produced under a Navy contract in 1917. This order called for nine of the rotund airships, and the first was completed on May 30, Decoration Day, 1917. It was the forerunner of what has become one of the world's best-known corporate symbols.

To produce these air vehicles, the company erected a hangar at Wingfoot Lake on the eastern edge of Akron. It was on a tract of 720 acres of land, which included the hangar, workshops, a hydrogen plant, and a landing field. Military cadets from MIT and Cornell University were assigned to Akron for lighter-than-air flight instruction at the lake under Goodyear experts. More than 600 men were trained there during the war.

## Wartime Problems

As the conflict progressed, the supply of rubber became a national problem. Rubber companies were allowed all the rubber they needed for war materials and government supplies, but restrictions were imposed on other uses. In the summer of 1918, automobile tire production was cut to half the output for the same period a year earlier. Goodyear discontinued bicycle and carriage tires and rubber bands in favor of war products.

The war also brought on labor problems. About 6,200 employees volunteered or were drafted, an unusually high percentage of total payroll because Goodyear had so many young employees. By the end of the war, more than 3,000 women worked in the Akron plant. Some even built tires and were paid the same rates as men. College students worked in the factory during the summer.

Early in the war, Goodyear set up special classes to train deaf-mutes for factory work. More than 250 entered the program and qualified as factory employees; about 40 were graduates of Gallaudet College in Washington, D.C., a school specializing in the education of deaf-mutes. The company's policy of hiring qualified deaf-mutes for production work continued for several years after the war.

Wages went up as the war effort progressed. The average hourly rate for labor in January 1917 was $0.464. It had jumped to $0.684 by November 1918, and on the first of that month an overtime program was established — time and a half for time worked in excess of 48 hours per week and double time for Sunday and holiday work.

Goodyear employees strongly supported the Liberty Loan campaigns and the Red Cross and War Chest drives. Ninety-eight percent contributed to the first Red Cross solicitation, furnishing one half of Akron's total quota. More than 11,000 bought the first Liberty Loan bonds issued, and 18,484 subscribed $541,316 to the War Chest, a healthy average in excess of $29 per employee, equal to five days' pay for many Goodyearites.

## Postwar Prosperity

When the war ended in November 1918, U.S. business and industry were moving at new high speeds, and Goodyear, the leading supplier to the exploding automotive industry, was in the forefront.

Goodyearites returning from military service found jobs waiting for them. For many, promotions and wage increases came quickly, and why not? Their company was speeding ahead to a promising future.

Sales had jumped from about $110 million in 1916-17 to nearly $172 million in 1918-19, and they reached $223 million in 1920. Goodyear supplied almost 50 percent of the original tire equipment for the U.S. auto industry.

It also supplied 60 percent of the new pneumatics and 35 percent of solid-tire requirements for trucks, 50 percent of the motorcycle tire market, and 60 percent of the needs of all carriage and buggy manufacturers still in business. General Motors was then Goodyear's biggest original equipment customer, reflecting Frank Seiberling's close relation with W. C. Durant, chief executive of that giant corporation.

Replacement tire sales boomed along with the rest, thanks greatly to all-out advertising and promotion that closely related to the success in auto racing. The straight-side cords got a strong promotional boost in 1916 when the nation's two best-known racing drivers, Ralph de Palma and Eddie Rickenbacker, set speed records with them, most frequently at the Daytona Beach track.

After a lapse during the war, the company again jumped into the racing arena. In 1919, de Palma drove to new records on cord tires at Daytona Beach, and in the Indianapolis 500 Mile Race, 27 of 33 starters rode on Goodyear straight cords, including the first- and second-place finishers, Howard Wilcox and Eddie Hearn. Thirteen of the 14 finishers had Goodyear cords.

*Top executives and managers, 1914. F. A. Seiberling, president and general manager (front row, 3rd from right); P. W. Litchfield, factory manager (middle, with boutonierre); G. M. Stadelman, secretary in charge of sales (last row, right). Seiberling was Goodyear's president from 1906-21, Stadelman from 1923-26, and Litchfield from 1926-40.*

The company withdrew from racing in 1919, satisfied that it had fully demonstrated the durability and safety of its straight-side cord tire in tough and highly visible usage. Racing, it then felt, contributed little to the design and development of road tires.

## City of Opportunity

The immediate future looked bright indeed to Goodyearites in 1919. More than 30,000 Goodyear employees in Akron averaged a healthy $1 an hour in wages. A second tire plant, a new six-story engineering building, and Goodyear Hall joined the corporate complex. The new Los Angeles factory was nearly completed, and the factory in Canada continued to flourish.

The Arizona cotton plantation showed $1 million annual profits. Export trade topped $8 million, and plans were brewing for a tire factory in Brazil. More than 15,000 gasoline service stations handled Goodyear tires, twice the prewar total, and $1 million in sales for each working day seemed just around the corner.

Still, Goodyear and its Akron competitors had an urgent problem. The city had been drained of its unemployed — workers were at a premium. Recruiters searched West Virginia, Kentucky, Tennessee, Georgia, and Alabama for factory help, hired them at once, and sent them to Akron on special trains. The greatest influx again came from West Virginia, Ohio's southern neighbor.

This population boom caused sudden problems. "Akron — Standing Room Only" is how one national magazine labeled it. Some workers took turns sleeping in the same rented rooms; a three-shift bed cost $5 a week per shift. Goodyear built barracks for its new employees.

Akron's rubber company employment zoomed from 22,000 in 1913 to 70,000 by 1920. The Chamber of Commerce erected a large billboard at Union Depot: "Akron, the City of Opportunity." And it was. By 1919, most rubber workers sported a silk shirt and often a felt beaver hat as signs of affluence. This was Akron's "Silk Shirt Era."

## The Industrial Assembly

In 1919, Goodyear took a long step forward in industrial relations. Under P. W. Litchfield's guidance, the company instituted the Goodyear Industrial Assembly to give employees a voice in management.

The assembly resulted from a study conducted by a 16-member committee, 8 appointed by management and 8 elected by employees. The committee chose a plan modeled on the U.S. Constitution with an assembly to legislate on all matters affecting employees, including wages and grievances.

It had two houses: a Senate of 20 members elected for two-year terms and a House of Representatives with 40 members elected for one-year terms. The assembly plan was approved by the board of directors and 90 percent of employees eligible to vote.

The plan divided the factory roster into 40 precincts, following department lines so that assembly candidates would be known by those they represented. Two precincts made up a senatorial district.

Standing committees were provided for wages, working conditions, safety, and welfare. Although the assembly could legislate on all matters affecting employees, it did not have final power. The factory manager and, if necessary, the board of directors could exercise a veto.

The Industrial Assembly Plan remained in effect for 16 years. It handled thousands of grievances, improved working conditions and safety, instituted hospitalization insurance, promoted a minimum wage, and obtained wage increases in 1923 and 1933 after extremely depressed periods. In 1924, an employee vote approved its continuance by 82 percent.

In his book *The Industrial Republic,* published in 1946, Litchfield wrote this about the assembly.

> The Industrial Assembly sought to establish a system of checks and balances to reconcile conflicting interests in industry such as the Constitution had set up in politics. It was started without fanfare, did not attract much attention at the time, and has been overshadowed since by other developments. It was not perfect, for the program would be carried out by ordinary people in management and the factory; but it was operated by men who believed in it, and who tried honestly and effectively to make it work. And through sixteen years of prosperity and depression it *did* work, and brought peace, understanding, and cooperation, and much better than average dollar gains to all concerned . . . and the two-way check on working conditions throughout the plant had highly beneficial results.

The assembly never received much acclaim from the business community, which considered it radical. However it did work for nearly two decades, and close associates of Litchfield said he considered it his most significant achievement.

From 1920 until its demise in 1937, the assembly met in Goodyear Hall, which was dedicated on April 20 that year by Ohio State University President W. O. Thompson, who said: "All Ohio is watching with interest this experiment in industrial education."

Inaugural speakers, including C. W. Seiberling and Litchfield, were introduced by Frank Seiberling. In his speech, Litchfield called the hall "a monument to the fact that hereafter, Goodyear intends to devote its energies to the building of men instead of the building of the machine." The first non-Goodyear speaker in the gymnasium was Senator Warren G. Harding of Ohio, who in a few months was elected the nation's 29th president.

In addition to the largest gymnasium in Ohio, the hall contained a theater that seated 1,786; classrooms, laboratories, and library; meeting rooms for employee clubs; bowling lanes; billiard rooms; rifle range; and a cafeteria equipped to serve 2,000 meals an hour. The fourth floor was reserved for women and had lockers, showers, a community room, a domestic science room, and other facilities for women's social and recreational activities.

Interest was so great that during the week following the inauguration, a crew of 50 guides was needed to show the new installation to employees and visitors.

The immediate postwar period was indeed the best of times for Goodyear. By the end of 1920, though, bright expectations were dimmed by the black clouds of economic disaster.

## A Postwar Scramble

Immediately after the 1918 Armistice, the federal government rushed to rid itself of wartime jurisdictions. It decontrolled prices only two days into peace and within a month canceled all its contracts.

U.S. industry also was in a rush — expanding as fast as it could to meet peacetime demands for goods no longer restricted. Prices rose rapidly, and by mid-1920 the wholesale price index registered 100 percent above 1914 and 25 percent higher than at the end of the war. Expectations of continuing price increases led to hurried accumulations of inventories, particularly of consumer goods.

Wages kept pace with prices, but productivity did not. Business and industry went overboard to obtain capital, regardless of high interest rates, and the industrial community scrambled to replace stock that had been exhausted during the war.

Suddenly the bottom fell out as business reached the point of diminishing returns: wholesale prices fell 33 percent, industrial production declined 25 percent, and unemployment rose to five million. Nearly 100,000 businesses went bankrupt, and in the next few years more than 450,000 farmers lost their properties. Overnight this recession had turned into the quickest and one of the most devastating in U.S. history.

For Goodyear the agony was even more acute. In his annual report to shareholders in 1918, F. A. Seiberling wrote these positive words.

> One year ago the Company had in excess of $15,000,000 Notes Payable outstanding. In view of the changed financial situation growing out of the war, and anticipating that the war might continue for several years, your Directors deemed it wise to increase the fixed capital of the Company, which was successfully accomplished by the sale of $15,000,000 of 8 percent Second Preferred stock to over 16,000 stockholders composed almost entirely of customers and employees — liquidating thereby its entire account of Notes Payable. There never has been a time in the history of the Company when its condition was as sound financially.

That was the first attempt to finance the growing capital requirements by direct sale to employees and customers without recourse to New York bankers, whom Seiberling distrusted. But the company's financial strength was ebbing.

Cash flow and working capital soon became huge problems. In 1919, authorized capital had been increased to $200 million, equally divided between preferred and common stock. In May 1920, Goodyear's directors authorized the proportion of two shares of preferred, one of common, and declared a 150 percent common stock dividend.

## Recession Nears

During Goodyear's two-and-one campaign, stockholders took little of the offering, but many employees borrowed from banks to get in on the deal, and the sales department sold stock in the field. The directors subscribed for $546,000 worth to express confidence in the company's future, paying with a joint note.

Sales in 1920 reached $192 million, and profits topped $51 million. A year later, sales had fallen to $105 million with a $5 million loss. In April 1920, production of pneumatic auto tires reached 837,236. By December, it had plummeted to 117,865.

About this time, *Triangle*, a company publication for the sales force, reported what gains would have accrued from a 1910 investment in Goodyear common stock if the holder had retained all dividends and exercised all rights to buy common offered to shareholders. A total investment of $6,064 would have amassed shares of $36,000 par value and cash dividends of $12,884.

That look at the past, however, contained no augury for the future, which would be an altogether different state of affairs — so different, in fact, that it would rock the very base of the Goodyear empire.

## Recession and Depression

The recession had started. Goodyear's cash reserves were low compared with the huge working capital required. Because neither Goodyear nor the Seiberlings had any Wall Street banking friends, financing was difficult to come by.

Costs continued high, inventories remained higher, and sales dropped. Notes and current accounts payable accumulated rapidly. Cash virtually disappeared. Sales were below estimates in April 1920 and failed to reach estimates for every following month through 1920. By year's end, they were only about half of the six-month estimate for July through December.

Goodyear consumed eight million pounds of rubber a month at the start of 1920; because of a delivery time from the Far East that ranged from two to four months, a constant six-month supply of 48 million pounds had to be maintained at a market price of 55¢ a pound. Fabric requirements were three million pounds per month at $1.50 a pound, and a workable inventory was a three-month supply.

The recession deepened. Rubber brokers, fabric mills, equipment manufacturers, and other Goodyear suppliers wanted their money. Banks recalled their loans.

Temporary refinancing of $18 million from a banking syndicate headed by Goldman, Sachs and Co., a New York house Goodyear had managed to win over, and A. G. Becker of Chicago was not enough. Further, the Goldman, Sachs syndicate managed the company's financial affairs so tightly that each day storerooms could issue only enough materials for one day of manufacturing.

*Members of the Old Guard, 1916, and (left) its founder, P. W. Litchfield. The Old Guard was formed in 1910 to honor 13 Goodyearites, including F. A. Seiberling, who had joined the company before 1900. The Clock Tower, built in 1915, was known as the Old Guard Tower for several decades because the Old Guard held many of its meetings there.*

Goodyear had closed its 1918-19 fiscal year with an inventory of $35 million, which by April 1920 had risen to $62 million. But crude rubber consumption at the Akron plant dropped from 10,070,336 pounds in March 1920 to only 1,483,056 in December. Fabric consumption in that span plummeted from 4.3 million pounds per month to 523,000. Sudden layoffs dropped Goodyear's 1920 Akron employment from 34,000 to only 8,000.

Looking back to 1920, Edwin J. Thomas — who in 1916 started as a typist and by 1940 had risen to the presidency — recalled "the heartbreaks for so many people."

One laid-off Goodyearite had told him that unless he could retain his job, he would go out in front of the plant and lie on the streetcar tracks until run over. Thomas said he felt just as badly, and taking him by the arm, agreed to join him.

"Well, he didn't," Thomas said, "and neither did I. But unless you have been through depressions like that, you can't imagine the tragedy when the breadwinner comes home to his family and has to tell them he no longer has a job."

## Rumors of Bankruptcy

Rumors of impending doom constantly ran through the office and factory areas. One, apocryphal or not, reflected the gloomy attitude that had taken over the company. Management had reportedly stationed a lookout on the East Market Street overpass of the B&O Railroad two miles from company headquarters. His job was to spot any potential bearers of bankruptcy papers and race back with the news.

Although the depression did not fully set in until mid-1920, every month that year showed bank overdrafts ranging from $200,000 to more than $2 million. Cash never reached 10 percent of liabilities, and the October 31 balance sheet showed only $1.67 million cash against liabilities of $41 million. The book value of common stock, $75 million in early 1920, had been wiped out entirely.

Refinancing was essential.

F. A. Seiberling's historical reluctance to let Wall Street gain any control over the company handicapped Goodyear in its bid for new money. The company had no friends among New York bankers. In fact, Seiberling had trouble raising money from any source.

Goodyear chroniclers and Goodyearites who knew F. A. are unanimous in praising his vision, his dynamism, his salesmanship, and his capacity to inspire confidence. Many considered him a business genius. But after the debacle that reached full climax in 1921, just as many faulted his disregard for prudent financial operations and a tendency to follow dreams and ambitions without keeping his financial feet on the ground.

## A Happy Warrior

Seiberling had built and led Goodyear to a greatness even he could not have imagined. He and many of his associates had become wealthy in Goodyear's first two decades, and Goodyear employees and dealers had shared in the company's astounding growth. But the "Little Napoleon," as some affectionately called him, met his Waterloo in 1921, and the company he founded nearly disappeared.

Tenacity was among Seiberling's strongest attributes, as his career had proved. He demonstrated this in his vigorous fight to keep Goodyear alive through 1921, and his management team fought valiantly beside him.

The battle was grim, but F. A. always showed his best face and at least in appearance seemed a happy warrior. While in New York's financial district on one of his forays to gain financial support, he ran into Harvey Firestone, there for the same reason.

"Harvey," the undaunted Seiberling reportedly said, "you take that side of the street. I'm working this side."

At the 1920 New Year's Eve meeting of the Old Guard, the ebullience and confidence of earlier meetings was missing. Most of the talk was subdued, nostalgic, and desperately hopeful, but the Seiberling sense of humor was still evident.

As the old year wound down and midnight tolled, Frank said to his brother, "Well, Charley, there goes 1920."

"I'm damn glad to see it go," Charley shot back.

The new year would prove even worse for the brothers. It would not only bring them more woes, it would lead to the end of their long association with the company they had struggled so hard to build.

# The Goodyear Tire & Rubber Company

## PLAN OF READJUSTMENT
## OF
## DEBT AND CAPITALIZATION

---

### PRESENT DEBT (EXCLUSIVE OF INTEREST) *

Bank Debt:

| | | |
|---|---|---|
| Special Secured Loan pursuant to Agreement of November 3, 1920, between the Company and Waddill Catchings and others, as a Committee | $18,825,000 | |
| Other obligations for the most part secured or partially secured by rubber which will be used in the current operations of the Company | 13,954,000 | |
| Unsecured notes and commercial paper | 12,202,500 | |
| | | $44,981,500 |

Ot ... 830
Bi ... 000
M...

... 104
Ta ... 026
Pr ... 701
Ot ... 573
... 734
Ot ... 556
... 290

PR ... CH

Fo ... 740
Fo ... 000
Fo ... 763
Fo ... 000
... 503

| | |
|---|---|
| Company's estimate of depreciation of materials covered by commitments, not heretofore written off | $18,247,000 |

### APPROXIMATE PRESENT CAPITALIZATION*

| | |
|---|---|
| Seven Per Cent. Preferred Stock | $65,000,000 |
| Common Stock | 61,000,000 |

**Treating the capital stock of the Company as a liability at its par value, the estimated deficit on December 31, 1920, exclusive of loss upon commitments for merchandise not yet delivered, was approximately $24,400,000.**

---

*** This information has been furnished by the Company.**

*The management team that led Goodyear from the edge of bankruptcy in 1921 to world leadership in the tire and rubber industry in 1926. First row, left to right: C. Slusser, P. W. Litchfield, G. M. Stadelman, W. Stephens. Second row: W. State, C. A. Stillman, E. G. Wilmer, T. A. Linnane, P. E. H. Leroy. Back row: H. T. Gillen, E. G. Huguelot, H. D. Hoskin, E. F. Eggleston, C. H. Brook.*

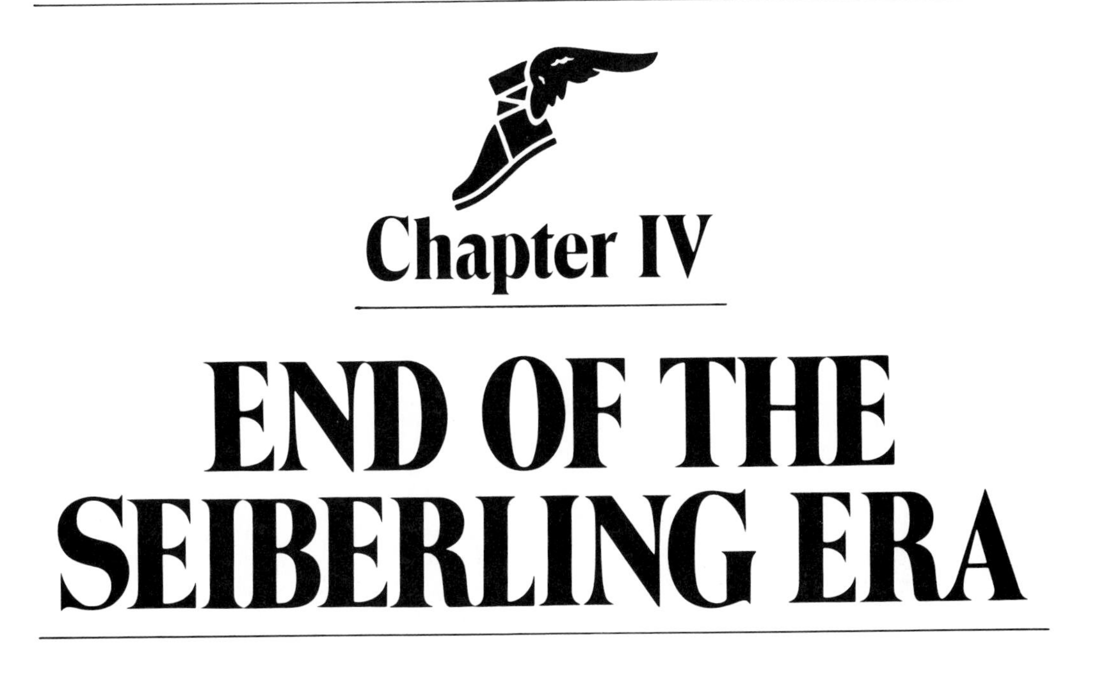

# Chapter IV

# END OF THE SEIBERLING ERA

Receivership loomed in early 1921. The New York law firm of Cravath, Henderson, Liffingwell & De Gersdorff, acting for Goodyear, set up a conference of creditors and bank representatives in January to discuss a rescue attempt. Several banks offered help if a responsible organization would direct the refinancing. Dillon, Read & Co. of New York, which had earlier worked with Goldman, Sachs on Goodyear finances, accepted the challenge.

Refinancing was a big-scale undertaking. It had to take up Goodyear's loans, pay off creditors, and provide working capital. Obligations exceeded $120 million, plus $18 million in losses on committed materials.

The bankers apparently had confidence in Goodyear's aggressive sales organization, manufacturing capabilities, commitment to innovation, and a national acceptance of the Goodyear name.

A plan submitted to the directors, which they approved, provided for recapitalization by issuance of new securities: eight percent bonds ($25 million); ten-year debentures ($25 million); eight percent prior preference stock ($35 million); and management stock (undetermined). A proxy form seeking shareholder approval was mailed; it also called for a special meeting in March.

The winter of 1920-21 had been an agonizing period for all Goodyearites. Many employees had borrowed heavily to buy Goodyear stocks and were being solicited for more collateral. C. W. loaned thousands of shares of his common holdings to help employees increase collateral and avoid being sold out. A $500,000 life insurance policy on F. A., payable to Goodyear, was canceled and its surrendered value paid into the company's treasury. Officers' salaries were cut 40 percent on April 1.

For months, company officers and the four refinancing committees labored to gain stockholder and creditor assent to the refinancing, which on May 11 was adopted without opposition.

The authorized capital structure, as indicated on the February 1 balance sheet and adjusted to show the effect of the financing, consisted of:

$30 million — first mortgage bonds;
$27.5 million — debentures;
$30.3 million — prior preference stock;
$65.5 million — seven percent preferred stock (old);
$10,000 — management stock;
$1 million — common stock.

To avoid a deficit on the asset side, the balance sheet carried an item of $12.5 million as "Special account, patents, and good will."

Syndicates bought the bonds and debentures at 90 percent of par; 275,000 shares of common stock were given with the debentures, one share with each $100 debenture. The bonds were redeemable at $120 and the debentures at $110, and both carried an interest rate of eight percent.

Prior preference stock was issued to assenting merchandise creditors at the rate of 125 percent of their claims. Dividends on that stock were to be at eight percent of par, which was $100, and were cumulative. The management stock trustees — Clarence Dillon of Dillon, Read; Cleveland banker John Sherwin; and Owen D. Young, chairman of General Electric Co. — would name the company's new directors.

The reorganization was completed on May 12, 1921, at a meeting of the outgoing directors in Dillon, Read's New York office. One of the chief purposes of this meeting was to receive the resignations of the board members, excepting Litchfield and Stadelman. F. A. presided as chairman of the meeting.

One by one the resignations were accepted, and each retiring director withdrew in his turn. Litchfield and H. B. Manton opposed the resignation of C. W. Seiberling; otherwise all resignations were unanimously accepted.

## Seiberling Steps Down

Frank A. Seiberling was the last to offer a resignation, which was accepted and seconded. The doughty, dynamic man who had founded Goodyear and led it both to preeminence in its industry and international renown, walked from the room, with poise and dignity.

Goodyear's Seiberling era had passed.

The May 17 issue of *Wingfoot Clan* closed its editorial this way.

> With the plan completed and in operation, Goodyear is out of the woods financially and can start clean.
>
> The period of doubt and uncertainty and uneasiness is past. We must face the future with unbroken courage. It is no time for regret over the set-backs and the sacrifices the past year has brought. Today is a new day. We render no service to the men who have left us by lamenting over their departure. Rather we must demonstrate anew their faith in us as men fit to carry on the great enterprise they founded.

A more emotional farewell in the same issue was written by Hugh Allen, author of a regular column in the *Clan*.

> Our respect and admiration for the Little Napoleon of a great industry have not grown less during these difficult months. For when a man fights a thing through, dogged, resourceful, unafraid, not asking quarter — fights to the finish, comes through at the end smiling, unbroken, undaunted, not asking sympathy, facing the world with his head up; well, it's an American trait to take off your hat to grit and backbone and courage. And besides all this, F. A. and C. W. were our friends.
>
> They built a mighty industry, far flung and worldwide; one that bulked big among the giants of business, but they never lost the human touch. They were our friends, friends to us of the shop.
>
> So today, while we can rejoice that Goodyear is to go ahead, that Litch and Stad [Paul Litchfield and George Stadelman] are to stay with us and sincerely extend to the new executive good wishes to success in his big responsibility, we'd be unworthy of the fellowship of the Clan of Goodyear if we did not pay tribute of honor and respect to our late leaders and let them know we were not unappreciative of their long friendship. God bless them.

Six months after his last Goodyear meeting, F. A. launched the Seiberling Rubber Co. with factories in Barberton, Ohio, and New Castle, Pennsylvania. It was the smallest of 300 rubber companies operating in the United States in 1921. Within a few years, it had grown to the eighth largest tire company in the nation.

In the darkest moments of 1921, many, including some of the company's top officers, thought Goodyear would collapse. Committed to saving the company, Litchfield and some of his lieutenants even studied the possibility of taking over a smaller tire company should worse come to worst. It did not, and the company's upward climb was directed by some of the same hands that had guided it to earlier heights — with help from the outside.

## Outsiders Move In

Goodyear's new chief executive was 38-year-old E. G. Wilmer. He had been associated with the Schlesinger conglomerate, headquartered in Milwaukee, Wisconsin, whose companies manufactured steel, coke, gas, and chemicals.

The management organization was supplied by Leonard Kennedy & Co., a group controlled by Dillon, Read and Schlesinger interests. The five-year contract between Goodyear and Kennedy was dated May 1, 1921, and could be terminated by owners of the management stock on two months' notice.

Nearly all the new management team came from various Schlesinger enterprises, and most had worked together there. Three Goodyear managers held their positions under Wilmer: George Stadelman, vice-president, sales; Paul Litchfield, vice-president, production; W. D. Shilts, assistant secretary. Two men from the accounting firm Price Waterhouse & Co., P. H. Hart and C. L. Weberg, who had been installed by the bankers in 1920 also survived.

Wilmer was cautious, testing his management team and relying heavily on Stadelman and Litchfield who had built recognizedly efficient sales and production organizations. The biggest changes were in the financial operations; a controller's division, new to Goodyear, was set up with assistance from Price Waterhouse.

The major criticisms of the Seiberling regime had centered on freewheeling financial operations, lack of controls, poor coordination between disbursements and income, and heavy contractual obligations undertaken with little regard for how to meet them. The budget system was weak. Taxes, real estate, and insurance were not centrally controlled. Overstaffing in administration and sales was rife, and the traffic department was considered inefficient. No centralized law department existed, nor did any legal control over contracts. The company had even been doing business in some states without licenses.

Capital expenditures for such investments as the Arizona land and Wingfoot Lake were considered excessive. The foreign branches were overstocked, overstaffed, and too indebted to Akron. Advertising expenditures of the past few years were termed extravagant.

Obviously in many ways the company had set itself on a sure path of self-destruction.

No one criticized the quality of Goodyear products or its sales operation. This was made especially clear immediately after Wilmer had assumed the top executive position and called Stadelman and Litchfield into his office.

"There is nothing wrong with Goodyear's manufacturing or sales," he told them. "The only problems are in finance. So I'll leave production and sales to you, and I'll see that you get the necessary funds."

Wilmer's positive attitude and restraint were exemplified in his initial statement to the organization, printed in the same *Clan* that bade farewell to the Seiberlings.

> I am fully satisfied with progress made [in the reorganization] and believe that Goodyear, with its finances adjusted, is ready for a complete comeback. Give credit to the men who have built up Goodyear — to the Seiberlings, and Litchfield and Stadelman, and the men they have gathered around them.

Perhaps the most difficult task for management was working out final settlements with creditors, but progress was rapid. On June 1, 1921, Wilmer reported that total debt (commitments excluded) had been reduced from $66 million to $26.5 million.

A September 30 report to stockholders indicated that sales of the parent company, excluding subsidiaries, for the past seven months had been $62.4 million; earnings before interest, miscellaneous charges, and adjustments were $6.8 million. Cash and its equivalents in the U.S. Treasury certificates and bank acceptances totaled $23.7 million. No bank indebtedness remained, and all trade accounts payable were being discounted.

Sales had picked up since spring, and 1921 employment by December was 8,100, up from 6,068 in March. In 1920, an average of .92 tire a day had been produced per factory employee, which in 1921 had more than doubled to 1.87. The average cost of labor per 100 pounds of output had dropped from $25 in 1920 to $11.70.

For the first time in its history, Goodyear had ample working capital, and of 469 creditor claims made in that year, all but 7 were completely settled.

A lot of fat still had to be cut from overall operations to get Goodyear in shape for the big climb back. "Retrenchment went into high gear," said Litchfield. Tire production had plunged from 30,000 a day to 5,000, and both the factory and sales forces had to be reduced, personnel cutbacks made with great skill and discretion. The vice-presidents intended to keep the best workers, giving preference to those with families.

## 'Let's Start Fighting'

In *Industrial Voyage,* Litchfield wrote:

> We could not have kept our best men if we had the union setup we now have; we would have had to lay them off in order of seniority regardless of ability. As it was we came out of the period with the most highly skilled men in the industry, and as business picked up, came back much faster than we could have done otherwise and all the sooner started calling men back to work.

Even before realignment, the branch operations staff had been halved. A thousand people were withdrawn from the general office operation, 700 salaried employees dismissed, and the sales staff cut in half. But the new top management wanted even more reductions.

"It was rugged going for a long time," Litchfield said.

> The morale of the organization went lower and lower . . . Many of our men were uncertain as to whether they would have a job the next day. I felt we were breaking the spirit of the very men needed to put the company back on its feet.

Convinced that a new attitude was necessary to get the company moving again, Litchfield vented his frustrations and enunciated his faith in Goodyear at a late 1921 meeting of branch managers, attended also by Wilmer and other top executives.

> Up to now this company has been retreating. That was necessary. But we've reached the point where we've got to quit retreating and start fighting. We've got to stop thinking about saving money and start thinking about making money. We can't pay off our debts by cutting expenses. We can only do that by going out after the business and getting it.
>
> We have plenty of weapons to fight with. The biggest one is our reputation. Our friends are sticking with us: General Motors, Ford, Nash, and the rest at Detroit, and our dealers all over the country. They have gone through the fire too. The public is for us. They have found that year after year Goodyear has given them the best damn tire in the world.
>
> This company was not built by ordinary men. It was built by thoroughbreds who outstripped their competitors in sales, production, development, engineering, and everything else. We can do it again. You men are still thoroughbreds. Let's start fighting.

Wilmer was first on the platform to congratulate Litchfield. "I am with you 100 percent," he said. All others in the room followed.

Sales picked up, and Stadelman and Litchfield were soon given the green light to start rebuilding the business. Many Goodyear executives of that time credited Litchfield's fiery talk with revitalizing management's attitude, turning the company around, and starting the comeback that in five years restored control to the stockholders.

## The Quarter-Billion Mark

According to Litchfield, "The only way we saw out of our troubles was through the same homespun formula we had always used, namely, to build better tires, cheaper, and sell them harder."

The formula worked — and it worked well. Sales climbed quickly from $103.5 million in 1921 to nearly $123 million in 1922. Stadelman was elected president in 1923, succeeding Wilmer who moved up to board chairmanship, and Litchfield was made first vice-president. Wilmer, concerned mainly with the company's finances and confident of the abilities of the Stadelman-Litchfield team, moved to Goodyear's New York office, taking the assistant comptroller and assistant secretary with him.

By 1926, the average mileage of a Goodyear tire was 12,000. Sales were $230,161,356 and would top $250 million by 1928. Goodyear's sixth president, however, who had led the sales force for a quarter of a century, did not live to see the quarter-billion-dollar mark. George Stadelman died on January 22, 1926, after a brief illness at the age of 53.

In a page-one obituary, the *Goodyear Triangle* pointed out that Stadelman "came to Goodyear when its sales were around a million dollars a year, and during his service sales passed the $200 million mark."

Although unassuming and reticent, Stadelman had had great confidence in Goodyear and its products and did not hesitate to express it. His confidence was usually at its peak on sales trips to Detroit, a lion's den for tire salesmen.

Several days before checking in at Detroit's Book-Cadillac Hotel, a favorite of his — and of the auto industry — Stadelman always placed an ad in the Detroit newspapers announcing his visit and that he would receive auto manufacturers wishing to discuss their tire requirements. Goodyear salesmen of later years, knowing the auto industry's tough and sometimes disdainful treatment of tire representatives, shook their heads in awe and envy at this lofty attitude.

## Litchfield: Chief Executive

At the annual meeting of the board on March 29, 1926, Litchfield was elected president; C. F. Stone, Cliff Slusser, and C. A. Stillman were elected vice-presidents; and W. D. Shilts became secretary.

A year earlier, a group headed by Dillon, Read had purchased Dodge Brothers, the auto manufacturer; Wilmer, still Goodyear's board chairman, became Dodge's president. On May 18, 1926, he resigned his Goodyear post, and Litchfield began what developed into a 30-year term as the chief executive officer. The company would function without a board chairman until he took over that responsibility in 1930.

In 1922, Goodyear's holders of common stock, represented on the board by a minority of five, began a long series of legal battles to improve the position of the common and preferred stocks and restore control to the stockholders. The Seiberling family and early directors were the prime movers of this effort.

Litchfield devoted most of his attention and energies in his first year as chief executive to resolving these disputes. The drawn-out litigation came to an end in 1927 after Owen D. Young, chairman of General Electric Co., was appointed mediator.

On May 14, the attorneys of record in all suits and the involved principals signed a "Memorandum of Agreement with Respect to Financing of The Goodyear Tire & Rubber Company and Termination of Stockholders' Litigation." The agreement's terms were overwhelmingly approved at a stockholder meeting on July 11. Litchfield was reelected president for three years and made a director, the only Goodyearite on the board of 17.

The company's common stock then sold at $50 a share; within six months it was $71 a share. The corporate capital was made up of $60 million in bonds, 650,000 shares of preferred stock, and 830,000 shares of common stock.

## World Leadership

Freed from backbending legal encumbrances, Goodyear was well-positioned for the market expansion of the late 1920s, indeed for the 10 years of trial and tribulation that would follow. In 1926, the world's largest tire company for the past decade also became the world's largest rubber company — measured both in sales volume and total factory output.

That year also marked an inauspicious move upward of a young man who had been working day and night for years to learn everything he could about the company, its jobs, and its problems.

Litchfield appointed Edwin J. Thomas, then 27, as director of personnel. It was the first executive position for the man who in 1940 would become Goodyear's eighth president and, 14 years later, chairman of the board. Thomas had begun work at Goodyear in 1916 two months before graduating from Akron's Central High School. He started part-time at $25 a month, but went on full time after graduation at $50. For several years he was a secretary in Litchfield's factory office.

He had made the most of opportunities there. He became a member of the Flying Squadron and learned as much as possible about engineering, processing, purchasing, costs, personnel, and other factory operations. For two years, Thomas worked in the factory each night on his own time and for no extra pay.

"I was young, healthy, and able to work," he said many years later.

> It occurred to me that if the company was willing, I could work in the factory after office hours and thereby learn all the operations of tire making. Mr. Litchfield readily agreed, and at the end of my office day I would go to the gatehouse and buy a box lunch, then change into my factory work clothes so I could work until 11 P.M.
>
> All of the men I met in the factory were very helpful, and I came to understand their problems and the things that bothered them; this helped me throughout my career.

Friendly, buoyant, and outgoing, Thomas participated in many company recreational activities and occasionally played the chimes in the clock tower. "The only problem," he recollected, "was that when you made a mistake, all of East Akron knew it. Then in the '30s, the Sunday chimes programs were stopped because of complaints from guests in the hotel across from the tower."

While an assistant to Litchfield, Thomas had been selected to help resurrect the Flying Squadron that had been discontinued during the hard times of 1920-21. He was to select 50 candidates and serve as squadron foreman in addition to his other duties.

In the squadron's first period, 1913 to 1920, nearly all candidates had come from the factory ranks. Thomas continued that system, but he also recruited directly from colleges and universities. This program, basically as he designed it, continues at Goodyear today.

To the veteran factory workers, the "college kids" were know-nothings who often got as much condescension as instruction. A grizzled foreman once asked a candidate, "Why do you think you should be hired?"

The young man proudly pulled out his degree. "Because I'm a college graduate."

"Look, Sonny," said the foreman. "You shouldn't keep a precious document like that all rolled up in your pocket. You should put it in a frame so heavy you can't carry the damn thing around. Then it wouldn't get in your way at work."

To instill confidence and team spirit in the newcomers, Thomas asked the trainees to vote on a squadron motto. They chose "When You're Green You're Growing and When You Get Ripe You're Rotten."

Although from 1922 the squadron included more and more college graduates, nearly a decade passed before many factory workers considered a college degree anything but a stigma. Until the 1950s and '60s, many of Goodyear's top production managers worldwide were not college educated.

## Export Business Grows

As the nation moved exuberantly into the advent of the Great Depression, Goodyear was on top of the world — almost. It was the world's largest tire company, since 1916; the world's largest rubber company, since 1926; and it was just beginning a worldwide expansion. In little more than 25 years, Goodyear had grown to a place among the titans of U.S. industry, and its international operations were growing rapidly. The annual report of 1928 described its far-flung operations abroad.

> Goodyear is now operating in 145 countries, with 174 distributors and 67 company-owned branches, depots, and warehouses, including 20 foreign selling subsidiaries. We have enjoyed an increasing proportion of the total potentialities of the export market. Our share of foreign replacement tire business in 1928 is estimated to have been 18.3% against 16.8% in 1927.

Despite great fluctuations in the size and scope of Goodyear, the company's character had changed little during its first 30 years. Management included people hired in before corporate annual sales reached a million dollars; its chief executive officer had been there since three years after the company began and had been its first factory superintendent.

The worldwide employee roster had grown from about 50 in 1898 to more than 40,000, but as some veterans said, "It still feels like a small company." Management remained personal, and Litchfield was a good example. Outwardly austere, he was accessible to everyone and was known as "Litch" to many lower-management members.

Guiding principles had not altered from an early determination to provide the best product and best service at the best price, to be the best company in the tire and rubber industry.

In *Industrial Voyage,* Litchfield wrote of his business philosophy when control went back to the stockholders and annual sales boosted Goodyear to the world's number one rubber company.

> We had gone through five years of banker management. They were able men, but the banking approach to business is necessarily different from that of the manufacturer.
>
> The banker thinks in terms of dollars and cents. That is his merchandise, his stock in trade . . . But the manufacturer is in a different position. He is dealing primarily with people, their wants, desires, ambitions. If he can produce something many people need, he will not merely get back interest on his money, but build a great business. In the endeavor to meet those needs, risks must be taken . . .
>
> A company can become a *big* company by doing these things. It may become a *great* company if it keeps service to society as its lodestar . . . A corporation, with its technical resources and financial strength, might become a *great* company if it used its position not for selfish purposes, but to expand still further its usefulness to society. That principle we must not forget.

## Visions of Greatness

This philosophy, which must have reflected some of the convictions Frank Seiberling had shared with Litchfield, had worked well for Goodyear in its first 30 years. Seiberling had visions of greatness for his company, and he took the risks necessary to make them reality. Almost from the first day of operations he dared all consequences in producing the Wing tire for carriages in the face of a patent infringement suit. He believed he was right — and he stood on his convictions. The profits from carriage tire sales had maintained the company for its first 10 years.

Seiberling foresaw the Automotive Age and began manufacturing automobile tires almost immediately. He sensed the clincher tire's shortcomings and pushed Goodyear's first primitive straight-side in 1900. He and Litchfield determined to make nothing but the best tires the company could produce.

Litchfield evidently shared Seiberling's belief in large-scale promotion, and by 1908 the two were intent on dominating tire advertising. Full-page Goodyear ads in the *Saturday Evening Post* and other national magazines were the tire industry's first.

Litchfield's experimental department of 1908 had developed such early product successes as the All-Weather tread, the cord tire, and for trucks, the pneumatic tire and the S V solid tire. By recruiting mechanical engineers to provide a broader base, he also departed from industry practice of using only chemical engineers in development areas. Further, his forward-thinking industrial relations policies instituted early in Goodyear's history were essential factors in the company's harmony with labor until the mid-1930s.

Other milestones in Goodyear's first three decades included a 1909 General Motors contract, which led to many more big sales in the auto industry. W. D. Shilts, who joined Goodyear in 1905 and was assistant secretary and secretary for 34 years, considered it perhaps "the most important order Goodyear ever obtained . . . It made Goodyear tires standard equipment on most General Motors cars for years thereafter."

The purchase of the Canadian factory in 1910 had started the company's global expansion. In the company's 30th year, the first two factories outside North America — tire plants at Sydney, Australia, and Wolverhampton, England — began production.

Marketing policies that successfully prevailed for many years had been established before 1910: Goodyear branches and salesmen sold original equipment; branches had the responsibility of wholesale distribution; and dealers, not jobbers or cut-price houses, conducted the replacement business.

Even more significant, perhaps, was the company's survival record during its first 30 years. In the face of both lean times and near bankruptcy, it had emerged strong and aggressive in a healthy recovery.

Its management was solid and seasoned, led by a chief executive who had proved his leadership and vision over nearly three decades. The management and shareholder confidence in Paul Litchfield was demonstrated convincingly two years later when, on March 31, 1930, he was reelected president and made chairman of the board of trustees.

Goodyearites of 1928 and stockholders had these reasons and many others to view the future with high hopes.

*A view from the ceiling of the Goodyear Airdock of the christening of the U.S.S. Akron, August 8, 1931. Mrs. Herbert Hoover christened the airship before a crowd of 150,000. Akron served the Navy until April 1933 when it plunged into the sea during a storm.*

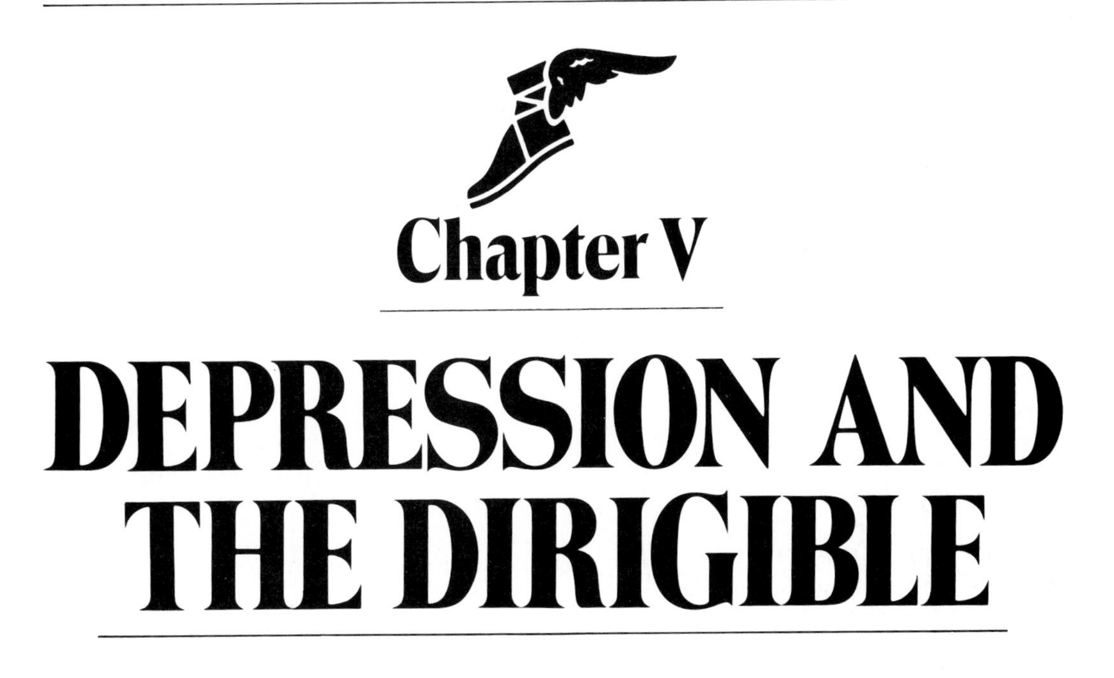

## Chapter V

# DEPRESSION AND THE DIRIGIBLE

Goodyear carried a distinctive corporate personality into its fourth decade. To some degree it reflected the characters of men who had led it to industry leadership and of employees who had helped shape the Goodyear family. This giant on the Little Cuyahoga was an entrepreneurial organization, innovative and aggressive, on the tracks to new horizons and ever-confident it would reach them.

Management still controlled tightly from on high, but the democratic and fraternal spirit of Akron and its agrarian roots had taken hold in corporate headquarters. Men who had worked their way to the top were on a first-name basis with many factory workers. Although sales in 1928 exceeded $250.7 million and Goodyear was the largest rubber company in the world, the family feeling remained strong.

It was enhanced by a comprehensive employee activities program with particular emphasis on sports, the employee publication *Wingfoot Clan* — best described as folksy — and a continuing reputation of management coming up through the ranks.

Early in 1929, Goodyear established its flag as a corporate symbol both at its facilities and on its products. The original blue and white banner of 1912 was redesigned by A. E. Boedeker, a well-known Ohio artist and company employee who had established the art department in 1915. It displayed the winged-foot logotype horizontally in white across a field of blue and gold.

Reminiscing in retirement a half century later, E. J. Thomas said, "I think it's safe to say most of our people were proud of what that flag represented."

Litchfield and his management team had helped keep the company alive in the early years, and they had weathered the reorganization of 1921 and the corporate upheaval that accompanied it. So when the company entered this new period, it had a thoroughly battle-hardened leadership crew, just what it needed.

For the world, the decade would be one of economic and political trouble and turmoil leading to the most catastrophic war in history.

For Goodyear, it would be a period of declining sales and profits and bloody labor strife. On the plus side, it would advance foreign expansion, great product and marketing innovation, and adventure in the skies. It also would solidify the company's character and personality through the exceptional foresight and leadership of the man many employees knew as Litch .

Two months after Goodyear's 31st birthday, the U. S. economy exploded its first frightening events of the most severe depression in the nation's history. The stock market crash of October 1929 signaled hard times that would continue until World War II.

The next 10 years held drastic changes in the world's power lineup and new directions for capitalistic democracy in the United States.

Worldwide, people would see Mussolini invade Ethiopia, Spain host civil war, an English king abdicate his throne, Germany repudiate the Versailles Treaty. Nationally, the U. S. government — not business and industry — would become the machinery to produce economic revival and social welfare.

Fifteen million Americans would be jobless in 1932. The major public programs and federal agencies of President Franklin D. Roosevelt's New Deal would become well-known acronyms to the man in the street.

NIRA (National Industrial Recovery Act)
NLRB (National Labor Relations Board)
PWA (Public Works Administration)
WPA (Works Progress Administration)
SEC (Securities & Exchange Commission)
AAA (Agricultural Adjustment Act)

The nation's banking system would collapse. Congress would enact the Social Security Act and the Wagner Labor Relations Act.

Goodyear would be in the thick of it all.

Being poor was not a social disgrace for Americans of the 1930s; if the great majority of Americans were not poor, they felt poor. In 1932, for the second time in the nation's history, an army invaded the capital city. It wasn't a foreign enemy, but a throng of 25,000 World War I veterans that stormed Washington seeking economic relief in immediate payment of the bonus government had promised them for 1945.

Toughened by the reorganization of the 1920s, Goodyear reacted quickly to the tearing changes already under way. Because of sound and innovative management, growth was steady in many ways, even though sales volume and profit both faltered.

It opened tire factories in such diverse areas as Gadsden, Alabama; São Paulo, Brazil; Buenos Aires, Argentina; Bogor, Indonesia; and Norrköping, Sweden. New textile mills started up in Georgia and Alabama, and plantations were developed in Indonesia, Costa Rica, and the Philippines.

The company expanded nontire product lines by introducing Pliofilm, a packaging material; Airfoam, a cushioning material; and Pliolite, a paint base and waterproofing compound. A plant in Windsor, Vermont, manufactured shoe products.

Goodyear's financial results roller coastered erratically through this period. By 1932, sales had plummeted to $109 million as the company operated at a net loss of $850,394, the first since 1921. Only three years earlier, 1929 had recorded profits of $18.6 million.

## The Big Airships

With sound financial underpinning, industry leadership, excellent public acceptance of products, and a farsighted, experienced management, Goodyear was positioned, however, for both survival and growth in the hard times ahead.

One avenue of growth, which would close tragically in the late '30s, was opened by P. W. Litchfield's vision of transoceanic passenger-carrying airships that would also serve as freight carriers.

In World War I, Germany's Zeppelin airships had demonstrated an ability to carry heavy loads over long distances and to fly well in poor visibility. Blimps had been used successfully by the Allies in observation and patrol duties, particularly by France and the United States.

After the war, several nations maintained interest in developing the dirigible. In the United States, which had a bountiful supply of helium, a nonflammable gas with about half the lift of highly flammable hydrogen, interest was especially high. Helium is seven times lighter than air, hydrogen fourteen times lighter.

In 1919, the U.S. Navy started work on the ZR-1, christened the U.S.S. *Shenandoah*, which in many ways was a copy of the Zeppelin L-49 design. The ZR-1 made its first flight on September 4, 1923. The Navy commissioned Germany's Zeppelin works, headed by Hugo Eckener, to build a dirigible in 1921 designated the LZ-126, later named the U.S.S. *Los Angeles*. This 658-foot-long airship was constructed in Germany and flown to Lakehurst, New Jersey, from the Zeppelin works at Friedrichshafen in October 1924. The 5,000-mile trip was accomplished in 81 hours.

The Navy liked the *Los Angeles* and hoped to acquire the technology and ability to build dirigibles better than the Zeppelin ships. In 1922, it asked Goodyear, which had built nearly every significant U.S. airship since 1911, to attempt the procurement of Zeppelin rights.

Litchfield agreed. He and Wilmer, Goodyear's president, traveled to Germany and arranged a deal. It provided for the transferral of Zeppelin construction rights to a new Goodyear subsidiary, the Goodyear-Zeppelin Corp., in which the Germans would have one-third financial interest. No money changed hands.

The deal was completed in 1924 when Zeppelin turned over its patents.

In 1926, Congress authorized construction of two dirigibles, but did not appropriate funds for them until 1928. In September 1928, Germany completed the *Graf Zeppelin*, which carried out nine years of continuous service. In all, this airship transported 13,110 passengers and traveled more than a million miles in 590 flights, including 144 ocean crossings.

*Right, Dr. Hugo Eckener (left), commander of the Graf Zeppelin and Lt. Commander Charles E. Rosendahl of the Akron. Below, ring laying ceremony inside Airdock, 1929.*

Eager to get experts to implement airship technology in the United States, Litchfield hoped to persuade Dr. Karl Arnstein, who had designed most of Germany's wartime Zeppelins, to come to Akron. Arnstein was a young Czechoslovakian engineer brought to Friedrichshafen at the start of World War I by airship designer Count Ferdinand von Zeppelin to help design airship hangars. He quickly became the leading engineering authority on airship construction.

Litchfield convinced the Zeppelin organization that Arnstein and a dozen engineers should move to Akron as part of the Goodyear-Zeppelin agreement, and the group arrived in 1925. Arnstein was a key figure in the company's lighter-than-air activities for the next 32 years and had an important management-engineering role in aircraft production during World War II.

In the fall of 1928, the Navy signed a contract for Goodyear to build two dirigibles. The designer would be Arnstein. The airships would become the world-famous *Akron* and *Macon*.

A crowd of 150,000 spectators, the largest ever assembled in Akron up to that time, attended the christening; network radio announcers described the scene, and newsreel cameras whirled as Mrs. Hoover pulled a cord to release 48 white pigeons, representing the nation's 48 states.

After several test flights in northeast Ohio, *Akron* was delivered to the Navy at Lakehurst on October 27, 1931. On its first flight as a naval vessel, the airship created a sensation when it cruised over New York, Philadelphia, Baltimore, Washington, and other high-population Eastern Seaboard cities.

The U.S.S. *Macon* was christened on March 11, 1933, by Mrs. William A. Moffett, wife of Rear Admiral Moffett, chief of the Navy's Bureau of Aeronautics.

Until *Macon* first sailed majestically into the night shortly after its christening — nearly two years after completion of *Akron* — the Akron airport attracted millions of sightseers and tourists during the four-and-one-half years after the hangar's ground breaking. Weekend crowds of up to 30,000 people regularly visited there.

## The Goodyear Airdock

Construction of the Goodyear Airdock for dirigible assembly began almost immediately. It was built at the Akron Municipal Airport by the American Bridge Co. from a plan by Wilbur Watson & Associates, architects and engineers of Cleveland.

The Airdock, a huge cocoonlike hangar, was the world's largest building without interior supports. It is 22 stories high with 364,000 square feet of floor space. Nearly 1,200 feet long and 325 feet wide, its eight-and-one-half acres could contain four football fields.

Identical doors at each end of the building are in two pieces and shaped like one eighth of an orange peel. Each section weighs 600 tons, moves on 40 wheels set on curved standard railroad tracks, and takes about five minutes to open or close.

Ground was broken for the Airdock in November 1928, and the U.S.S. *Akron* was christened by Mrs. Herbert Hoover, wife of the president, on August 8, 1931.

*Above, a Navy plane grabs the trapeze of the airship U.S.S. Macon. In a hangar built into its huge hull, the Macon could hold five planes that could be launched or recovered in flight. Opposite, the U.S.S. Akron beside Hangar One at the Naval Air Station in Lakehurst, New Jersey, 1932. It was on a flight from Lakehurst the following year that the Akron flew into a storm and crashed into the sea off the New Jersey coast.*

Public interest in construction of the two airships was stimulated by dirigible activity elsewhere. In August and September of 1929, the *Graf Zeppelin*, commanded by Eckener, made an around-the-world flight in 20 days and 4 hours. The U.S.S. *Los Angeles* toured the United States, demonstrating in-flight launching and pickup of airplanes.

In August 1930, Great Britain's rigid airship, the R-100, made a round-trip flight between England and Canada, generating a new wave of interest in the potential of rigid airships for transoceanic passenger service.

## Transoceanic Dirigibles

During the construction years of *Akron* and *Macon*, two companies — International Zeppelin Transport Corp. and Pacific Zeppelin Transport Co. — were formed primarily to investigate routes, bases, and schedules for transoceanic transportation of passengers, mail, and cargo. Litchfield, an intense believer in the airship's viability for commercial travel and freight haulage, was in the thick of the planning as chairman of the board of Pacific Zeppelin and president of International.

He indicates in *Industrial Voyage* that little difficulty was encountered in wooing investors.

> The steamship people, the airlines, industries like the Aluminum Company, Carbon and Carbide, General Motors, banking houses like National City, G. M. P. Murphy, and Lehman Brothers, and California and Hawaii business houses took a similar interest in the projected transpacific line. Goodyear-Zeppelin was to be the constructor. People in the shipping business, with experience in that field, would operate the ships.

According to early plans, the transatlantic schedules called for 2½ days from Washington to Paris and, because of headwinds, 6¼ for the return. The Pacific route would start in California for Hawaii, Manila, Japan, and possibly China. This trip should require 6½ days — 36 hours from California to Hawaii, 72 more to Manila, then 48 to Japan. The return trip would probably take from 4 to 6 days, including a stop for mail and passengers in Honolulu.

Legislative roadblocks in Washington were heavy, almost immovable, and long before they could be pushed aside, Litchfield's dream ended with the gradual, tragic finale of the era of airships.

On October 5, 1930, the five million-cubic-foot capacity British airship R-101, on a nonstop flight from England to India, crashed into a fog-shrouded hilltop near Beauvais, France, killing 48 of the 54 people aboard. The British government quickly abandoned further participation in rigid airship activity and scrapped the R-100, sister ship of the R-101, after an uneventful two years of operation.

The *Akron* plunged into the Atlantic Ocean on April 4, 1933, during a thunderstorm 25 miles southeast of Barnegat Inlet on the New Jersey coast. Only 3 of the 76 men aboard survived, and among those lost was Rear Admiral Moffett, whose wife had recently christened the *Macon*, which a week later began its trial flights.

Scouting large areas of ocean at speeds of up to 80 miles an hour, the *Macon* operated nearly two years for the Navy without incident. Five planes could be housed in the hangar built into the bottom of the airship's hull. These craft were launched and recovered in flight with a device called the trapeze.

Tragedy eventually overtook the *Macon* as well when in February 1935 it took part in exercises with the U.S. Pacific Fleet off California. After maneuvers on the 12th, the *Macon* set course for its home base, Moffett Field, near the south end of San Francisco Bay.

Nearing the coast, it was hit by a maelstrom of air currents while working its way through low clouds, rain squalls, and fog. The crew battled the storm for a half hour, but all directional control was eventually lost and the *Macon* hit the water. In another 30 minutes, the ship had broken up and been swallowed by the Pacific.

Thanks to an SOS the *Macon* had sent out before crashing, 79 of the crew of 81 survived, picked up by ships of the Pacific Fleet that had steamed to the rescue. The *Macon* was the last rigid airship to fly under the American flag.

The glory years of the airship ended in tragic finality on May 6, 1937, when the German Zeppelin *Hindenburg*, completing a scheduled transatlantic crossing at Lakehurst, burst into flames just seconds before landing. The *Hindenburg* was the largest commercial aircraft ever flown (seven million cubic feet) and in 1936 completed 10 transatlantic trips.

This rigid airship, like the *Graf Zeppelin*, was inflated with volatile and sensitive hydrogen, not the helium that was used in U. S. airships. In all, the *Graf Zeppelin* and *Hindenburg* had carried 20,389 people and 283,067 pounds of freight and mail in 634 trips that covered 1,200,385 miles.

On its last flight, the *Hindenburg* had 38 passengers and a crew of 59. Thirty-five failed to survive the disaster — 13 passengers and 22 crewmen — plus one member of the ground crew.

No aerial incident had received such vivid, on-the-scene news coverage as this one. Before hundreds of people — newspaper and news service reporters, newsreel and other photographers, radio broadcasters, naval personnel, friends and relatives of those aboard, and residents of nearby communities — the airship era died out with the flames of the *Hindenburg*.

For Paul Litchfield, the demise of the big rigid craft was almost certainly the greatest disappointment of his life. From the very beginning of aviation, he had been commercially and aesthetically interested in these giant airships and in aviation in general.

Goodyear had built its first airplane tires in 1909, and in 1910 Litchfield visited England and France to learn more about airplanes and their performances in air meets there. He had used this new knowledge in the redesigning of Goodyear airplane tires.

In 1954 he wrote:

> With a conservative New England background and a hardheaded engineering training, I do not think anyone would write me off as a harebrained dreamer. I got into the field [airships] because I believed, and still believe, that the passenger-carrying airship is practical from a construction and operating standpoint, and has great possibilities in international commerce and trade . . . Two or three reconnaissance airships, able to patrol the ocean from Alaska to Panama, might have prevented the tragedy of Pearl Harbor.

*Opposite, construction of the airship U.S.S. Macon inside the Goodyear Airdock. Construction of the Airdock began in 1928, and when completed, the giant hangar was the world's largest building without interior supports. Right, famed aviatrix Amelia Earhart (2nd from right) christens the Defender at the National Air Races in Cleveland, 1929. Defender was the largest airship of the Goodyear fleet.*

## Goodyear's Blimp Fleet

As the big rigid airships made headlines in the 1930s, their small sisters, the blimps, were being greatly improved and before long would get even more exposure than the dirigibles.

Both blimps and dirigibles — sometimes called Zeppelins — are airships. Dirigibles and Zeppelins have rigid internal frameworks. Blimps are nonrigid; the shape of a blimp's envelope — the bag that holds the helium — is maintained by the internal pressure of the gas, which lifts it from the ground.

The word blimp is supposed to have been coined by Lt. A. D. Cunningham of Britain's Royal Navy Air Service in 1915. During an inspection of the Capel Air Station in England, Cunningham playfully flipped his thumb on the side of a nonrigid airship. When an odd noise echoed off the taut fabric, Cunningham imitated the sound: "Blimp!"

With the construction of four small nonrigid airships, Goodyear's blimp operations had expanded to fleet proportions by 1929. These new blimps supplemented the promotion and public service activities of the *Pilgrim*, built in 1925, and the *Puritan*, 1928. They were named after U.S. contenders in the America's Cup international yachting races, a custom that has continued.

*Volunteer* went to the West Coast, and the *Mayflower* and *Vigilant* toured East Coast and Midwest cities. Similar to the *Puritan* and slightly larger than the *Pilgrim*, these early airships carried four passengers and the pilot. The later blimps, somewhat larger, carried six passengers. The *Defender*, largest of the Goodyear fleet, was christened at the National Air Races in Cleveland on August 30, 1929, by the aviatrix Amelia Earhart.

*Defender's* envelope had a capacity of 178,000 cubic feet, and the cabin comfortably accommodated eight passengers. Initially, *Defender* operated from Akron, but its home base was a new hangar adjacent to the Gadsden factory, a plant also constructed and opened that year.

Most blimp improvements and the methods of handling them were the result of pragmatic experience by the Navy, the Army Air Corps, and the wide-ranging Goodyear fleet. The Air Corps maintained its blimp program until 1935 when because of a lack of funds it turned over all blimp equipment to the Navy.

The high point of Air Corps blimp development was the TC-3 of 1933, the largest nonrigid airship built up to that time. It was a 350,000-cubic-foot airship with a cloud car, a streamlined capsule that could be lowered several hundred feet by cable to permit its single passenger to reconnoiter the ground or sea at close range while clouds concealed the blimp.

Although the Navy purchased and experimented with several TC-type blimps in the 200,000-cubic-foot category, its most significant research was conducted with Goodyear's K-1, which had a capacity of 320,000 cubic feet. Research in antisubmarine warfare conducted with the K-1 resulted in a new class of naval patrol airships. Goodyear built the prototype of this new class, the K-2, in 1940; its capacity was 404,000 cubic feet. A total of 135 model K ships were built, more than any other kind of blimp.

The Goodyear fleet expanded operations in the '30s as blimps established themselves as a Goodyear hallmark throughout the nation. *Columbia* and *Reliance* joined the team in 1931, *Resolute* in 1932. The Goodyear fleet operated constantly, and its blimps accented such events as the Chicago World's Fair, the New Orleans Mardi Gras, Cleveland's Great Lakes Exposition, the Rose Bowl football game, the National Air Races, and the 1932 Olympics in Los Angeles.

## Altitude Record Set

Goodyear blimps and Goodyear-made dirigibles operated near the earth's surface, but fabric balloons the company made were part of spectacular tests to determine how high man could fly. In 1933, the Goodyear-made *Century of Progress* airship, sponsored by the Chicago World's Fair, the *Chicago Daily News*, and the National Broadcasting Co., attempted new altitude records.

The first effort, cheered by 50,000 people at takeoff from Chicago's Soldier Field, was a spectacular disappointment. Mechanical problems brought the big 600,000-cubic-foot balloon to earth just a few miles away. In November of the same year, though, the *Century of Progress* lifted off from Akron Municipal Airport and carried Navy Lt. Comdr. T.G.W. Settle and Marine Maj. Chester Fordney to a new altitude record of 61,237 feet. Settle and Fordney were the first humans to ascend more than 10 miles above the earth's surface.

In 1934, the Army Air Corps teamed with the National Geographic Society for further exploration of the stratosphere, and Goodyear manufactured the huge test balloons. In the first test, *Explorer I* climbed to more than 60,000 feet. *Explorer II* — its capacity of 3.7 million cubic feet equivalent to the *Graf Zeppelin* — carried a crew of three to a record 72,395 feet in four-and-one-half hours.

While involved in lighter-than-air aeronautics, the company was keeping pace with development of the airplane. Contributions centered on wheels, tires, and brakes and by World War II had established Goodyear as the runaway leader in their design and manufacture. Many speed, distance, and transcontinental aviation records set in the 1930s by such noted fliers as Roscoe Turner, Jimmy Doolittle, Howard Hughes, and Wiley Post were made by planes equipped with Goodyear wheels, tires, and brakes.

*Jimmy Doolittle at Mitchell Field near Hempstead, Long Island, New York, 1928. The renowned aviator set many of his speed and distance records in planes equipped with Goodyear wheels, tires, and brakes.*

Among the company's spectacular achievements in aeronautics was the Goodyear Airwheel, introduced in July 1929. This revolutionary tire had a full balloon casing mounted directly on a hub that eliminated the conventional rim, spoke, and wheel system. It provided large air volume at low pressure and had the advantages of better cushioning for landing, improved traction in snow and on ice and wet ground, and better ground gripping in crosswinds.

The company purchased two airplanes for promoting and testing the Airwheel. One, a Fokker Super Universal, visited most airports and well-known flying fields in the country on an extensive barnstorming tour. Veteran airmail pilot Clarence Bell was at the controls, and L. O. Guinther, Goodyear's manager of the airplane tire and accessories division, went along as the tire expert and salesman. The Fokker had dual controls, and qualified pilots were invited to land the plane to experience the full effect of improvements the Airwheel provided.

Goodyear introduced the first hydraulic disc brakes developed expressly for airplanes in 1933. Until then, brakes of this kind for airplanes had been mere adaptations of automobile brakes. Goodyear's new brake applied the principle of a metal disc clutch and had a power input efficiency rating of 97 percent, compared with 35 percent for the average mechanically operated brake, which soon obsolesced as planes became larger and more powerful and carried heavier loads.

In 1937, Goodyear engineers developed pneumatic brakes for the large bombers and big commercial planes that were equipped with tires four to five feet high.

## Black Monday Looms

Encouraged by record sales of $250 million in 1928, Goodyear's directors voted to resume payments of dividends on common stock in August 1929, just two months before Black Monday, October 29, signaled the beginning of the Great Depression.

Dividends continued until the end of 1932. They were resumed in 1937, but suspended within a year when the national economy dipped and sales fell sharply. Payments resumed in the first quarter of 1939 and through 1982 have continued without interruption.

Despite a sound financial structure, experienced management, and good public acceptance, Goodyear, typical of U. S. industry as a whole, was dramatically shaken in the 1930s by the depression and ensuing changes in economic philosophies, industrial relations, and government influence on business and industry.

The debilitating effects of this most serious of all depressions in the nation's history continued for many years. Short periods of partial recovery emerged, but the economy did not regain its pre-1929 momentum until World War II accelerated the wheels of industry.

*Crowds of strikers and union supporters mill outside Goodyear's general office entrance on February 24, 1936. The five-week strike ended one month later, but work stoppages and violence continued into the following year and led to the end of the Industrial Assembly.*

# Chapter VI

# THE HEARTBREAK YEARS

For Akron and the tire and rubber industry, the future had seemed bright and rosy in 1929. Akron's Big Three broke all sales records that year: Goodyear's $256 million was followed by Goodrich's $164.4 million and Firestone's $144.5 million. New buildings filled downtown, and the city grew up at a furious pace.

Depression in the following year — both economic and psychological — quickly changed the atmosphere in the Rubber Capital of the World. As elsewhere, middle-class families suddenly were poor; the chorus of a song popularized by crooner Rudy Vallee had a poignancy that struck home with many Akron businessmen.

> Once I built a railroad, made it run,
> Made it run against time.
> Once I built a railroad, now it's done.
> Brother, can you spare a dime?

In October 1930, Akron's City Council appropriated $75,000 to clear underbrush at the water reservoir in Geauga County. More than 900 men applied for the 60 available jobs: eight hours a day for 50¢ an hour.

The Citizens Unemployment Committee provided apples to sell on the streets, as similar organizations did in many cities, and more than 700 people applied for the 200 jobs.

From the time of the stock-market crash to the end of 1930, Akron factories discharged 14,200 persons, an employment drop of 21.5 percent.

To shore up the shaky economy, the federal government enacted dramatic legislation under Roosevelt's New Deal program. The National Recovery Administration (NRA), headed by Gen. Hugh Johnson, was the New Deal's standard-bearer to revive and strengthen industry. Roosevelt promoted the NRA in fireside chats, and Johnson, an expert showman, made support of the program a national crusade. NRA parades were held throughout the country, and the news media, Hollywood promoters, and business in general went all out to support this program, which would "rescue us from the depression," as Johnson promised.

Its basic purpose was to eliminate cutthroat competition and unfair labor practices, particularly those involving women and children. Companies that abided by the NRA code displayed the NRA emblem, a blue eagle over the legend "We Do Our Part," and President Roosevelt ordered government agencies to do business only with NRA-related companies.

The business community, however, gradually came to view NRA as a new form of bureaucratic control that restrained competition and turned against it. Organized labor elements were jittery about NRA, fearing that its industrywide agreements and price-fixing elements could smother their unions.

Despite the hoopla with which it was born and the national fervor that greeted it, the NRA died a slow death as business, labor, and the public all revolted against its dictatorial powers, its centralization, its encouragement of price fixing, and its control of the economy. In 1935, the U.S. Supreme Court ruled the NRA unconstitutional.

The agency had provided strong impetus to labor. It made collective bargaining legal, and because it laid the groundwork for employees to organize into unions without employer interference, the NRA had been widely interpreted to imply that all workers should join unions.

When the court declared the agency illegal, its collective bargaining provisions were incorporated into the Wagner Act of 1935, which established the NLRB and provided even stronger guarantees for organized labor. Although the NLRB's legality was challenged, the Supreme Court sustained it in April 1937.

Ironically, Goodyear, with its long record of pioneering in industrial relations, became a battleground for an acid test of the New Deal's labor relations philosophies.

## A Good Labor Record

Employee relations at Goodyear had been good — far more stable than in U.S. industry as a whole — since the company first opened its doors.

In 1913, Goodyear had experienced its first labor problem when factory personnel joined a strike that involved an estimated 15,000 workers at the Big Three. None of the plants closed down, but production, conducted by skeleton crews, was minimal.

The strike had been motivated mainly by Industrial Workers of the World (IWW), an extreme left-wing labor union; local Socialist leaders were among the prime movers.

It was unpopular, and at the end of its second week a citizen welfare league was organized to help keep order in the streets. On one occasion, a large contingent of league members marched from downtown to the Goodyear factory gates in support of nonstrikers.

The Ohio State Senate sent a committee to study the situation, including Senator William Green who soon would head the American Federation of Labor. The committee reported that rubber workers earned an average of $2.60 a day, compared with a national labor average of $1.40, and that working conditions were good.

By a lopsided vote of the workers on March 30, the strike was abandoned after 48 days. But it did stimulate gradual gains for Akron's factory employees.

The following year, Goodyear started the eight-hour workday and adopted a vacation schedule for factory workers: a week with pay for those with 5 to 9 years of service and twice that for those with 10 or more. In 1916, a new pension system enabled long-service employees to retire on income based on earnings.

The Employee Activities program, a pet project of Litchfield and the Seiberling brothers, was begun well before World War I and swung into high gear with the completion of Goodyear Hall in 1920. It included employee teams in baseball, bowling, soccer, tennis, golf, and riflery and clubs for employees interested in fishing, hunting, and other outdoor sports. The Christmas party for employees' children annually drew 25,000 youngsters, and the annual picnic at Euclid Beach on Lake Erie for employees and their families was attracting 50,000 by the late 1920s.

Industrial sports programs had grown increasingly popular during the 1920s, and Akron's rubber companies were natural rivals in intercompany competition. Goodyear's employee athletics program grew into one of the most extensive in the country during this time as thousands of Goodyearites participated in company-sponsored activities.

On July 14, 1929, Goodyear dedicated a new brick and concrete stadium at Seiberling Field, and more than 8,000 people witnessed a dedication baseball game with Firestone. After the Goodyear band played the national anthem, Paul Litchfield threw some pitches to Harvey Firestone to open the new facility.

Following World War I, basketball had become the most popular sport at Goodyear and over the years retained that edge in popularity. Early coaches were Edwin "Chief" Conner, Edgar "Smiley" Weltner, and Clifton "Lefty" Byers. Each turned out top-notch teams, and the 1936-37 squad won the Midwest Conference Championship. The 1938-39 team won the championship of the National Basketball League, the strongest basketball organization in the nation and forerunner of the National Basketball Association.

*Above, Harvey Firestone (left) and P. W. Litchfield at the opening of the new stadium at Seiberling Field, 1929. Right, pickets opposite Goodyear State Bank, February 24, 1936.*

A star of that team was Victor Holt, Jr., an all-American from the University of Oklahoma who in 1964 became president of Goodyear. E. J. Thomas, who assumed the presidency in 1940, had been a member of the 1920 team.

## Labor Trouble Begins

In 1932, Goodyear established a short workday in Akron — six hours for each factory shift — to spread work, reduce layoffs, and keep as many employees as possible on the payroll. The plan had been presented by management to the Industrial Assembly, which submitted it to all employees. A large majority approved it.

When Goodyear Hall was completed in 1920, the company also erected a bank building in the triangle formed by the hall's west end and the intersection of East Market Street and Goodyear Boulevard. The building, opposite the company's main entrance, was occupied by the Ohio Savings & Trust Co. until President Roosevelt's national "bank holiday" of March 1933, when Ohio Savings merged with a downtown bank and left East Akron without a bank branch.

As a convenience for employees and East Akron residents and businesses, the company established the Goodyear State Bank in the vacated building. It furnished basic bank services, and its first president was Thomas O. McEldowney, who had been chief examiner for the Ohio state banking division.

Employee relations at Goodyear had appeared solid in 1930, but as the depression and the New Deal changed the very ethics and shape of labor relations, Goodyear was buffeted by winds of change.

Shortly after Congress passed the NIRA in 1933, the American Federation of Labor (AFL) began a drive to organize the rubber workers. One of the AFL's top organizers, Coleman Claherty, came to Akron in mid-1933 to direct the campaign and at the end of the year claimed an aggregate membership of 20,000 in local plants. About 1,000 workers signed with the Goodyear union.

Local union chapters demanded in 1935 that the rubber companies officially recognize them as bargaining agents for employees; the AFL local would replace Goodyear's Industrial Assembly. When the companies rejected these demands, the unions threatened a citywide strike. The Industrial Assembly arranged for a strike vote by secret ballot in which only production workers, no foremen or supervisors, could vote.

The result was 11,516 against striking and 891 in favor. Firestone and Goodrich conducted similar in-plant votes with similar results, but at a lower margin: approximately seven votes against a strike for every pro strike vote.

When the unions announced they would base strike action on a vote restricted to members of the locals, the federal government intervened. Secretary of Labor Frances Perkins held separate Akron meetings with company and union executives, and a settlement, called the Perkins Agreement, was eventually reached.

It provided that the companies would meet with employee representatives to discuss grievances, hours, wages, and working conditions and would post notices of any changes in these areas on factory bulletin boards. No elections to determine the bargaining agencies would be ordered by the National Labor Relations Board until the constitutionality of NIRA had been determined by the courts.

Until it was, no strikes or lockouts could occur, and any grievances not settled satisfactorily would be referred to a fact-finding board appointed by the Secretary of Labor. The threat of strike was averted as union activities — for a while — were lulled.

At the first international convention of the United Rubber Workers of America (URWA) in September 1935, an estimated total of 60 rubber worker locals were formed in the United States with about 4,000 members, compared with 30,000 a few years earlier.

When the Supreme Court nullified the National Industrial Recovery Act that year, NIRA-imposed price controls were removed. Price wars broke out that diluted rubber company profit margins to break-even points at best. With Goodyear's Akron wages 48 percent higher in 1936 than in 1932 and as much as 50 percent higher than competitors' elsewhere, management reestablished 8-hour shifts in some departments of Plant II, which varied work from 32 to 40 hours a week. The company said that resulting efficiencies would reduce the cost of tires, generate more business, and eventually provide more work.

## Strike Vote

In reaction to the revision of work schedules, the Industrial Assembly enacted a bill requesting the 6-hour shift as standard policy in Akron. Clifton C. Slusser, vice-president and factory manager, vetoed the bill, pointing out that the shift was designed to give all employees an average work week of 36 hours. The assembly appealed to the board of directors, but the veto held.

Local 2, encouraged by the assembly's unexpected militancy, appealed to the Secretary of Labor, alleging a violation of the Perkins Agreement, and a fact-finding committee was sent to Akron from Washington in late November. In January 1936, it ruled that Goodyear was unjustified in returning either wholly or partly to the eight-hour shift because the company had established the shorter workday on its own volition.

The committee also charged the company with "discriminating" between the Industrial Assembly and the union, even though the assembly had officially represented employees for 16 years and less than 10 percent were union members.

In essence, the Perkins Agreement was interpreted to mean the company could take no action on wages or hours without first consulting the union. Although the fact-finding committee had no authority to enforce its findings, Goodyear scheduled work at Plant II in six-hour shifts.

Nevertheless, tire markets shrank, sales dropped, inventories accumulated, and layoffs became necessary. On the Monday following a sit-down at Plant I on Friday, February 14, 1936, the union voted a strike and began picketing on the midnight shift at Plant II. By Wednesday noon, several thousand pickets huddled in the below-zero cold both at Plant I and the office building. They had built shelters of two-by-fours and canvas and had fires burning in metal drums outside the gates.

Goodyear's Akron factories were besieged by pickets for the next five weeks. Although only a few hundred were members of Local 2, the Goodyear unit of the URWA, they were supported by large groups from other rubber companies and other unions and unionists sent in by John L. Lewis and the Committee for Industrial Organization (CIO) from coal mines in Ohio and West Virginia.

*Wood shanty picket post at factory office entrance protected strikers from the bitter cold of the winter of 1936.*

More than 1,000 employees, including Litchfield and other officers, lived and worked in the plants throughout the strike to protect Goodyear's property and produce and ship whatever products they could. Their food was delivered by U.S. Postal Service trucks as picketers allowed only government vehicles to enter the gates. Even this service was stopped after two weeks, and deliveries could be made only by trains, manned mostly by management crews, which brought in food and carried out a few tires on each trip.

Court injunctions to restrict the number of pickets were ignored, and county and city police forces were inadequate to control the situation.

## Goodyear the Target

As labor strife wore on, the weather matched it in bitterness with frequent subzero temperatures and almost daily snow. Strikers built larger canvas and wood shanties, and flaming oil drums encircled the factory building.

The strike's basic purpose was to establish the union as the bargaining agency for all factory employees. Grievances were aired during discussions, and the piecework system was denounced as a "speed-up" device, but the final settlement concentrated on means by which employee representatives could have more say in factory affairs.

Goodyear's labor relations hit the national spotlight during the 34-day strike. The nation's labor movement, which considered this confrontation a test case in the "new deal" for unionism, supported the strikers completely.

Veteran union leaders from other industries invaded Akron to lend strength to the obvious all-out strategy: crack Goodyear, the largest rubber company, and destroy its history of amiable labor relations. With this accomplished, the other rubber companies could more easily be brought into line and organized by labor.

A thoroughly professional effort was thrown into the conflict, a concert of power that seemed to surpass all capabilities of recently organized locals headed by union newcomers. It included administration, mass meetings, organizing and manning the picket lines, directing and feeding the pickets, and public communications and press relations.

Goodyear principals during negotiations were E. J. Thomas and Fred W. Climer, who were accompanied by Lisle M. Buckingham, the company's local legal counsel, at meetings with union representatives.

READ THE ADS FOR BARGAINS THAT MEAN SAVINGS TO YOU

*The Paper That is Read in the Homes*

EAST AKRON NEWS

CALL HEMLOCK 4142 FOR NEWS AND ADS CALL HEMLOCK 4142

Vol. 1[illegible]—No. 38 AKRON, OHIO, THURSDAY, MARCH 19, 1936 PRICE 3 CENTS

# Strike Situation Tense

## GOODYEAR NOW IN SECOND MONTH OF LABOR TIEUP

### Reopening Of Plant Threatened By Management

*Left, the March 19, 1936, "East Akron News" headlined the story absorbing Akron, along with labor and business leaders across the country. A front-page editorial exhorted the union's president and Goodyear's president Litchfield to "stop making East Akron the battleground for a national fight between capital and labor."*

After two years as general superintendent of the Los Angeles plant, Thomas had returned to Akron as assistant factory manager. When William Stephens died in 1932, Thomas succeeded him as general superintendent of the Akron operations. In January 1935, he was named managing director of the Goodyear Tire & Rubber Co. (Great Britain) Ltd., headquartered at Wolverhampton, and returned to Akron about a year later as assistant to the president, less than three weeks before the strike.

Greeted by news of impending labor difficulties, Thomas visited the factory and talked things over with old friends. He asked them for a more cooperative and conciliatory attitude, but was told: "We can't, Eddie. It's a whole new ball game."

Climer had joined Goodyear in 1919 and served in the labor department until being named Akron personnel manager in 1928. When in that year the Argentine plant opened, he became its general manager until being recalled to Akron in 1934, again as personnel manager. He was made corporate director of personnel in 1938 and nine years later became vice-president for industrial relations.

On March 24, 1936, Goodyear and the union agreed to a one-page, seven-paragraph memorandum of settlement, published two days later over Litchfield's signature in the *Wingfoot Clan*. He pointed out in the editorial that the Industrial Assembly would occupy exactly the same position as in the past, emphasizing that management would continue to deal with all employees, regardless of affiliation.

The settlement restored production, but not normal operations. Work stoppages, mostly sitdowns, often occurred at the slightest provocations — even on apparent whims. Some workers who had resisted union organization or preferred to have the assembly represent them were beaten up or chased from the plant. Departments were sometimes shut down to protest the assignment of an antiunion employee.

Union leaders could not control members. In the first two months after the settlement, management faced 15 sit-downs ranging from brief department tie-ups to the complete 24-hour disruption of Plant II.

In an editorial of May 21, the *Akron Beacon Journal* expressed its alarm at the situation.

> The epidemic of disruptive, outlaw sit-downs in the rubber industry has reached the point of crisis. It will have to be stamped out quickly if the economic life-blood of Akron is not to be destroyed . . . Sit-downs, as they have been used in Akron, are guerrilla warfare — undeclared, ruthless, uncontrollable.

*Mob assembles before Willard St. gate, February 25, 1936.*

## Industrial Assembly's End

On May 23, the *Beacon* again editorialized:

> An incident occurred at The Goodyear Tire & Rubber Company yesterday that unfortunately threatens to put the stamp of gangster terrorism upon the union labor movement in the rubber industry. Were Akronites to read of a similar occurrence in a factory in Germany, in Russia, or Italy, they would either be outraged or wholly incredulous. In the very heart of Plant II, with scores of workers nearby, a group of 25 men fell upon and seriously injured two fellow employees.

The wave of disruption and violence did not subside quickly. In the ensuing year, more than 150 sit-downs occurred in the company's Akron factories.

In July 1936, the URWA formally affiliated with John L. Lewis's CIO, which had been formed as an instrument of the AFL in November 1935. The CIO was suspended by the AFL in September 1936, but the URWA maintained its loyalty to the Lewis faction. On April 12, 1937, the Supreme Court validated the Wagner Act (National Labor Relations Act), and two weeks later Goodyear's Industrial Assembly was disbanded.

The company had provided no financial support to the assembly during its early history or to members of its two governing houses. After a few years, though, some members, particularly officers and committee chairmen, complained of losing money because their assembly duties interfered with their regular work. The company reluctantly approved the organization's proposals to be paid for time spent on assembly affairs at regular pay rates.

The Industrial Assembly was therefore in the "financial support" or "company union" category under terms of the Wagner Act, thereby assuring its dissolution.

At the company's request, on August 31, 1937, the National Labor Relations Board conducted an election on the issue. The question: "Do you want to be represented for collective bargaining by Goodyear Local 2, United Rubber Workers of America?" Of the 13,021 workers eligible to vote, 8,464 voted yes, 3,192 no.

The demise of the Industrial Assembly evoked cheers from the union, but unauthorized work stoppages continued. They climaxed on May 26, 1938, with a sit-down in Plant I that brought pickets to the gates even though no formal vote had called for them. City police went to the plant ordering the pickets to disperse.

A melee broke out. Tear gas, clubs, and stones were used, and more than 100 men were injured, none seriously.

Three days later, the company and union negotiated an agreement, normal work schedules resumed, and sit-downs decreased. The riot apparently had had a sobering influence on wildcatters, putting a damper on the sudden work interruptions that had become bothersome both to Goodyear and the union.

Although labor peace had returned to the rubber companies in Akron, the strife of the 1930s added a stimulus to build up tire production in other sections of the nation. Akron's 1939 rubber company employment was 33,285, compared with 58,316 in 1929.

After three years of harmonious and responsible labor relations, Goodyear entered into its first formal contract with Local 2 at Akron in October 1941.

When the time came to sign, Litchfield told Thomas and Climer he could not put his signature to the contract. Although he recognized that the Industrial Assembly — one of his proudest achievements — could never return, he felt a great sense of loss and a certain disappointment in his factory boys, as he called them, whose team he had led for so long.

"I understand your feelings, Litch," Thomas said, "especially in view of your pioneering leadership in industrial relations. However, I know you appreciate that we must live with the union, and that the new situation will not change our management attitude toward our people. We will never lose confidence in them."

Thomas signed in Litchfield's place.

## Expansion Goes On

Despite the tribulations of the Great Depression, Goodyear added a great deal of production capacity in its fourth decade. It started in the South, at Gadsden, Alabama, where a new Goodyear tire factory saw its first tire built on June 22, 1929. In Washington, on July 11, President Herbert Hoover had pressed a button, sparking an electrical impulse that traveled over telegraph wires to Gadsden, releasing the Goodyear house flag on the factory tower and officially opening the new plant.

Original capacity was 5,000 tires a day.

That year two fabric mills were added south of the Mason-Dixon line, both in Georgia: one at Rockmart, the other at Cartersville. The Rockmart plant, built from the ground up, was the company's first completely new textile mill. The facility in Cartersville was acquired from American Textile Co.

Goodyear acquired another mill, the fourth in its chain of fabric manufacturing plants, in late 1933. Built in 1927 by the Connecticut Mills Co. in Decatur, Alabama, it was equipped to produce tire fabrics.

The company's first major acquisition of a complete business organization was in 1935 with the purchase of the name, goodwill, and assets of the Kelly-Springfield Tire Co. This was, of course, the same company that Goodyear had challenged in the courts years earlier.

Kelly-Springfield was about as old as the U. S. auto industry, although it did not assume its double name until 1914. In 1894, Edwin S. Kelly and Arthur W. Grant had established the Rubber Tire Wheel Co. at Springfield, Ohio, to produce a new solid tire for carriages. Kelly provided most of the financial support and Grant the mechanical know-how.

In his Springfield blacksmith shop, Grant had developed a method of drawing two strands of wire imbedded in a strip of rubber around a carriage wheel to cushion the ride and eliminate clatter on cobblestone streets. The product was patented and named Kelly-Springfield. Goodyear had been involved in a series of bitter lawsuits over these Grant patents.

In 1899, Rubber Tire Wheel was separated into two companies: the Buckeye Rubber Co., for manufacturing, and the Consolidated Rubber Co., for sales. These companies built a factory in Akron south of Goodyear's Plant I. On January 2, 1914, they were reorganized and merged into the Kelly-Springfield Tire Co. This manufacturer was a major factor in the replacement tire market, but never a prime competitor in the battle to sell to the automobile manufacturers.

Kelly-Springfield had relocated its headquarters from New York to Cumberland, Maryland, in 1921, and a phaseout of its Akron operations had begun. Hit hard by the depressed replacement tire market and squeezed by fierce competition, the company went into bankruptcy in 1935 under the new National Bankruptcy Act.

Later that year, Goodyear acquired it as a separate subsidiary with its own manufacturing and sales organizations. Edmund S. Burke, who 24 years earlier had joined Kelly-Springfield as a clerk and worked his way up to the presidency, was retained in that position. He was made chairman of the board of Kelly-Springfield in 1959 and retired in 1960.

Goodyear wanted a plant convenient to Detroit's concentration of motor vehicle manufacturers and in 1936 purchased a group of factory buildings at Jackson, Michigan. This complex — constructed by Kelsey-Hayes Wheel Co. and operated for only a few years — was in six months converted to a tire plant with a capacity of 5,000 tires daily.

In that year, the company also bought an idle machine tool factory in Windsor, Vermont, for shoe sole and heel production.

For three major reasons, operations in shoe

products, established in 1900, were shaky, even though Goodyear maintained industry leadership.

1) Most of the big shoe manufacturers were in New England, remote from Akron. That increased shipping costs and made relations difficult.
2) Wages in Akron were much higher than those paid to shoe factory workers in New England. This exerted a strong upward pressure on prices in competitive markets.
3) Shoe products required only a small fraction of the rubber consumed in Akron, most of which went into tires. For this reason, they were almost always second in line for special compounds to meet specifications of shoe manufacturers.

The dilemma was resolved and the number one industry position maintained by purchase of the Windsor plant. It was converted quickly to shoe products and on November 16 produced its first heels.

More than four decades later, E. J. Thomas reminisced that this plant "might have been named the Morse-Post factory." He said that in the mid-1930s, Herman Morse, in charge of shoe products development, and Harry Post, director of sales, argued for shifting production from Akron to the Vermont site. They were told they could, but to "make your own decisions on its operation, and if it succeeds, the plant will bear the Goodyear name. If not, it will be the Morse-Post factory." The operation was successful from the start.

## Foreign Markets Beckon

As automotive markets developed abroad, so did the export appetites of U. S. tire manufacturers. Goodyear's first export orders had been from Great Britain at the turn of the century when the English import firm of Daws & Allen arranged to distribute straight-side tires. Exports to other transatlantic countries followed on a small scale, and until just before World War I, Goodyear's shipments from Canada, mainly to Great Britain, were greater than those from the United States because exports from the "sterling countries" enjoyed preferential tariffs.

After establishing a branch in London in 1912, Goodyear made plans to secure a foothold in the European market. L. C. Van Bever, an executive of the Canadian subsidiary and first manager of the London sales branch, traveled the Continent, establishing distributors in Germany, Russia, the Netherlands, Austria, Sweden, Denmark, and Finland. Sales agencies were also arranged that year in the Caribbean, Central America, and South America.

When Wilmer became president in 1921, he brought in a new export manager, Frank K. Espenhain, who soon separated export from domestic operations. The Goodyear Tire and Rubber Export Co. was established in 1922 to expand international business. A. G. Cameron, who had come to Goodyear from grocery wholesaling in 1913 as a truck tire salesman, was made manager of the new subsidiary, which for many years was known within the company as "Export."

The Canadian subsidiary, manufacturer of most of Goodyear's exports in the 1920s, increased its sales in 1922 by 684 percent over 1921. This trend continued in varying degrees, and as the '20s closed, Goodyear was on the verge of becoming multinational on a large scale.

For several decades, management had been eying the potential tire markets of South America. Litchfield had visited Brazil in 1912 and considered it a nation of great promise. Between then and the late '20s, plans had been outlined frequently for tire plants in South American countries, but none was implemented until 1930, when Goodyear announced it would construct a tire factory at Hurlingham, a Buenos Aires suburb. This would be the fifth foreign plant, following those in Canada — Bowmanville and New Toronto — England, and Australia.

The Hurlingham site had been selected before the depression, and because of Argentina's potential, management decided to construct the factory despite adverse business conditions at home. Within a few months of its opening on January 24, 1931, the Hurlingham factory was in full production of 1,000 tires a day.

Far to the north, the oldest foreign subsidiary, Goodyear Canada, had become firmly established in the Dominion's industrial community. Seven years after its founding in 1910 it was Canada's largest tire manufacturer. Among those who led it to the top was C. H. Carlisle who had come to Goodyear in 1908 from Akron's M. O'Neil department store.

Carlisle had been sent to Toronto by F. A. Seiberling in 1909 to scout the first Canadian Motor Show. He returned greatly impressed with Canada and its people and, mainly because of his enthusiasm, was promptly sent back to open a Toronto branch.

When Goodyear Canada, Inc., was founded a year later, Carlisle was its secretary-treasurer. He soon became a Canadian citizen and was a key figure in the management team that produced an average profit of more than $1 million a year from 1910 to 1940. He was Goodyear Canada's president from 1926 to 1936 and later became president of the Dominion Bank, one of Canada's largest.

Plans to establish Goodyear's ninth tire factory — its sixth outside the United States — were announced in August 1934; it would be at Bogor, Java, Indonesia. Among a group of highly trained personnel sent there to get the plant constructed and equipped was Russell DeYoung, who was responsible for the installation and operation of the tube manufacturing department.

DeYoung had received the Litchfield Medal as outstanding production squadron graduate of 1933 and when chosen for Indonesia was a tube-room supervisor in Akron. He would become Goodyear's chairman of the board and chief executive officer in 1964.

## Foothold in East Asia

The Bogor plant produced its first tire on April 29, 1935.

Two new plantations also were established in the 1930s, emphasizing the company's drive to reduce dependency on Southeast Asia for natural rubber supplies. The first, primarily an experimental station, was in Panama. Negotiations with the Panamanian government to buy 2,850 acres on the shores of Gatun Lake were completed in 1935.

Called the All-Weather Estate, this new plantation had many varieties of high-yield rubber plants that had been transplanted from Sumatra and the Philippines for cultivation and study. In 1943, it was turned over to the Inter-American Institute of Agricultural Sciences, a research and educational organization sponsored by member countries of the Pan American Union.

A second experimental plantation, in Costa Rica, was acquired in 1936. Consisting of only 1,000 acres, the Speedway Estate was soon expanded to 2,500 acres and became the principal rubber plant research center in Central America.

W. E. "Moe" Klippert, who later became vice-president and general manager of the rubber plantations division after a long career in the

plantations, was then manager of plantations in Central America and for several years supervised the work at the All-Weather and Speedway estates.

Two more overseas tire factories were producing before the end of the 1930s. Plans to construct a plant at Norrköping, Sweden, were announced in June 1938, and its first tire came off the line on February 23, 1939. The first managing director of Goodyear-Sweden was Thure Melander, a native of Sweden who came to Akron in 1922 to learn the tire business and later was sales supervisor for Scandinavia and Iceland.

When a South American tire factory was first considered — in 1912 — Litchfield had selected a tentative site in Brazil, but 26 years passed before a former textile mill in São Paulo was bought in November 1938 and converted to a tire factory. Remodeling began in December and production started on July 14, 1939.

By the time Adolf Hitler invaded Poland on September 1, 1939, the operations of the world's largest tire and rubber company were truly global. Goodyear had 18 subsidiaries, 7 factories, 7 plantations, 37 branches, 28 depots, and several hundred distributors outside the United States. Its products were sold on every continent except Antarctica and advertised in more than 20 languages.

A new subsidiary, Goodyear Foreign Operations, Inc., had been established that year to

*Opposite, tapping rubber tree, 1935. Goodyear's first plantations, established in the '30s, demonstrated a determination to reduce dependency on foreign rubber suppliers. Above, dealer delivery truck fashioned as replica of Goodyear tire, 1930.*

replace the Export Co. and manage all foreign operations except those in Canada; Litchfield was president and Thomas and P.E.H. Leroy vice-presidents. Leroy was a 19-year veteran in the company's financial operations and was elected a corporate vice-president in 1932. A. G. Cameron, who had been manager of the Export Co. since its beginning in 1922, was general manager and vice-president, and G. K. Hinshaw was vice-president, production.

Steady — sometimes drastic — changes had marked Goodyear's fourth decade. Despite the negative economic effect of the depression, however, most major changes represented progress and a strengthening of industry leadership.

International expansion had positioned the company extremely well as motor travel blossomed around the world. The airship adventures, though unprofitable, had attracted new and favorable public attention to Goodyear and its pioneering spirit. Labor unrest, searing in its effect, had provided early experience in emerging relations between U.S. labor and management. As resources in capital and technology had accumulated, technical progress had accelerated.

The advanced development of Goodyear's first commercial synthetic rubber, Chemigum, and a pilot plant that demonstrated its mass production capabilities rank high in the company's achievements of the '30s.

After Goodyear patented the basic processes for making Chemigum in 1927, natural rubber prices fell. This caused synthetic rubber development to lag until 1933 when Dr. Ray Dinsmore, assistant to the factory manager and responsible for new product development, initiated an extensive research program to manufacture synthetic rubber for tires and industrial products.

As the war clouds formed, so did a large-scale interest in synthetic rubber. In 1935, Dinsmore visited Germany to investigate progress on synthetics there and was shown many products made from a synthetic rubber called Buna-S, but its production techniques were closely guarded. What he saw had been helpful, though. He returned to Akron and set up a small plant that soon turned out 10 pounds a day of the first U.S.-made synthetic rubber to come off a production line — Chemigum.

Tires were produced with Chemigum in 1937, and Goodyear road tests demonstrated its superiority over the German synthetic. By 1939, the pilot plant's capacity had been increased to one ton a day, and a year later the first complete Chemigum production plant was built in Akron, with annual manufacturing capacity of 2,000 tons.

As Chemigum evolved, research scientists made other discoveries that greatly expanded Goodyear's business in chemical products and related fields. Pliolite was introduced in 1934 as a resinous substance for cementing rubber to metal. Research and refinements later made it valuable as a corrosion-resistant coating for metals, a base for waterproof paints, a film to waterproof paper, and, during World War II, an ingredient in bullet-sealing fuel tanks.

Pliofilm had also come out in 1934 as a transparent, thermoplastic film made by treating pale crepe natural rubber with hydrogen chloride. It later gained wide use as a packaging and preserving material for vegetables, meat, and other foodstuffs and as an element of rainproof garments and umbrellas.

Goodyear had made rubber tile for floor coverings as early as 1905, but this market was small until the mid-'30s when rubber flooring and wall coverings became popular. The company was ready with a tile much improved over earlier versions.

A continuous flow of other products were introduced in the 1930s.

- Airfoam, a cushioning material for mattresses, seat cushions, and upholstery
- Neolite, a material for shoe soles and heels of better durability than leather and rubber
- The first pneumatic tires for towed farm implements
- The first heavy-duty low-pressure tires for earth-moving vehicles
- The first studded "mud and snow" tires
- The Sure-Grip "go anywhere" tires for heavy vehicles
- Tires for use in cane or rice fields
- The first open-center, Sure-Grip farm tractor tires
- The first amphibious marsh buggy tires
- The first wide-base farm tractor tires
- Passenger car tires with rayon cord
- Hi-Rib tires for front wheels of farm tractors
- Low-stretch Supertwist tire cord

By World War I, tire manufacturing had become one of the leading industries in the United States. Goodyear, Firestone, U.S. Rubber, and Goodrich were known to most Americans, and as the automotive industry grew, the tire industry grew with it. During the 1920s, tire distribution expanded at a hectic pace; when the depression struck, 183,000 tire retailers of all kinds were doing business in the United States, more than double the number of just five years earlier.

Retailing expansion was due in great part to the oil industry and its thousands of gasoline service stations and to the proliferation of chain stores set up by such large mail order companies as Sears Roebuck and Montgomery Ward. The tire industry itself went into retailing on a national scale with the establishment of retail stores by the major rubber companies. These outlets provided distribution in areas inadequately supplied by the mass merchandisers or smaller stores.

*Allis Chalmers tractor, with Goodyear farm tires, operating in a hay field on an Illinois farm, 1933. Several designs of new farm tires were part of a flow of new products introduced by Goodyear in the 1930s.*

As the economy declined after 1929, redundancy in tire retailing became commonplace. Sales exceeded $200 million at Goodyear only twice through 1939, and its common stock dropped to $10 a share in March 1933 from $106 a share in October 1929. In the 1930s, Americans had more pressing matters on their minds than buying cars and tires, the major one being economic survival.

The signs and sounds of war, however, multiplied at a headlong pace across the Atlantic, bringing along a huge new customer and an exceptional new market for almost all of industry. This new client would be the government of the United States.

## The 40th Anniversary

As Goodyear entered its fifth decade on the eve of World War II, it continued its hold on global leadership in the tire and rubber industry, had gained solid manufacturing footholds in burgeoning foreign markets, and was still led by many of those who had guided it to the top.

In 1937, E. J. Thomas had been elected to the board and became the first executive vice-president. R. S. Wilson, a Goodyearite since 1912, was the top sales executive, elected vice-president and sales manager in 1928 and a director in 1932. Cliff Slusser had been in charge of production and a vice-president since 1926, a director since 1932.

P. E. H. Leroy had been elected vice-president, finance, in 1937. J. M. Linforth, a 20-year veteran in sales, was assistant to the president and would become vice-president in charge of original, aeronautic, and government sales. Fred Climer was director of personnel; Russell DeYoung was assistant to the president, concentrating on aircraft operations; R. P. Dinsmore was development manager; and Vic Holt, the all-American who had compiled an outstanding record in a variety of sales positions, was a sales manager.

In February 1939, seven months before World War II erupted, Goodyear celebrated both its 40th anniversary and the 100th anniversary of the discovery of the vulcanization of rubber by Charles Goodyear. The event centered on an Akron gathering of more than 1,800 key company personnel from around the globe.

Most of the world kept a nervous watch on Hitler's agitation for "living room" in Europe, and one million people had died in Spain's Civil War. Even so, the prospect of the United States again going to war seemed remote to Goodyearites at the start of 1939, especially in light of the Munich Peace Pact of September 1938, which allayed some of the concern about a major conflagration across the Atlantic.

Many Akronites were making plans to visit the New York World's Fair, local payrolls were expanding, and the rubber factories were putting up "Help Wanted" signs. The pinch of the depression was now barely a squeeze.

The 40th anniversary homecoming was a massive celebration with a series of pageants and presentations on the history of the industry and the growth of Goodyear. The company had its own "World Fair of Rubber" in Goodyear Hall where departments presented 50 exhibits of their products and services. The Seiberling brothers, now heading the Seiberling Rubber Co., were on hand for many of the activities.

One anniversary highlight was the unveiling of a statue of Charles Goodyear in a small park between the downtown National Guard Armory and the Summit County Courthouse; this statue remains an Akron attraction. C. W. Seiberling was master of ceremonies, and Mayor Lee D. Schroy accepted the statue as a gift to the city from Goodyear.

During his remarks, C. W. stumbled in pronouncing Goodyear. His brother later whispered to him, "You said Goodrich."

"The hell I did," Charley snapped back. "I'm a Goodyear man."

A decade later, in October 1948, the Golden Jubilee anniversary was marked by another reunion in Akron and a series of anniversary banquets in 54 U. S. cities and a dozen foreign sites where Goodyear sales or manufacturing facilities were headquartered.

The catastrophic events of the decades spanning these anniversaries, which shifted the global balance of power and changed the face of the world, transformed Goodyear from a maker of products for peaceful purposes into a mass manufacturer of the implements of war — and back again.

Nevertheless, through all the trauma of those changes, through the gains of war and often greater losses, Goodyear maintained the personality and character developed during its first 40 years. Most importantly, the corporation remained loyal to a couple of well-known principles that tenaciously remain as the foundation of Goodyear philosophy: "Protect Our Good Name" and "People Are Our Most Important Asset."

*Goodyear office employees leave Goodyear Theater after attending one of four "D-Day" prayer services, part of the observance on the home front of the beginning of the invasion of Axis-held Europe. Bulletins were read throughout the plant during the first days of the invasion.*

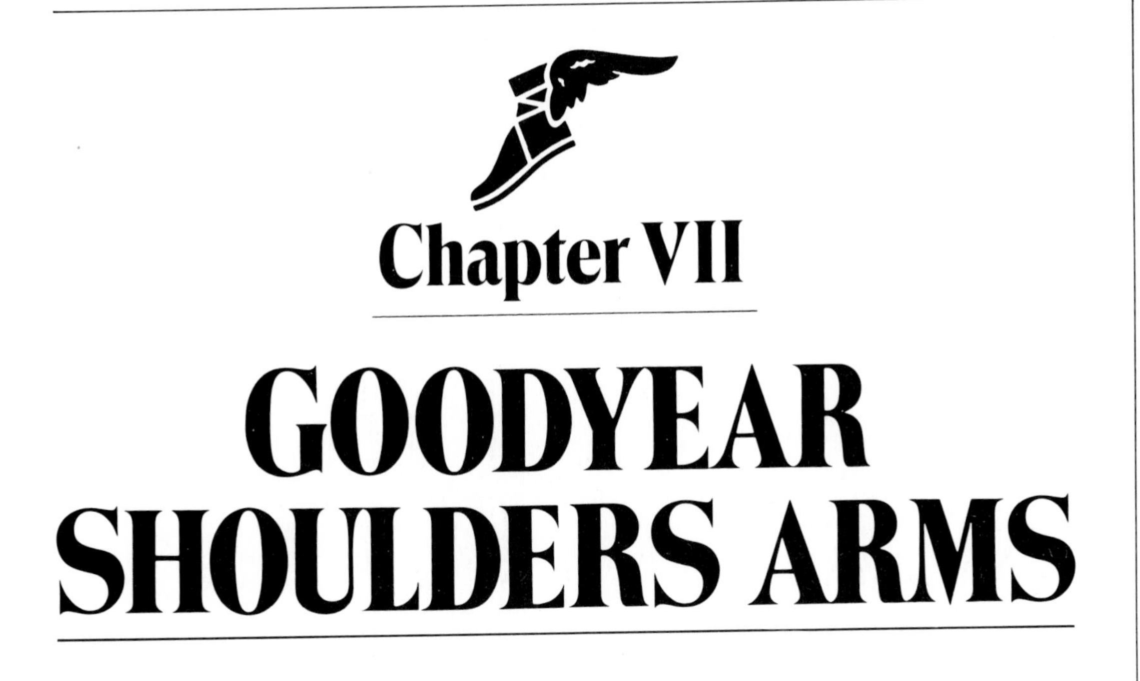

# Chapter VII

# GOODYEAR SHOULDERS ARMS

Like most American companies, Goodyear was quick to shoulder arms in the first industrial war, a worldwide conflict with an outcome determined mainly by technology and machines. Its contributions and activities were many.

- 4,008 *Corsair* fighter planes — called Whistling Death by its Japanese foes — for the Navy
- Wheels, brakes, and many additional parts for many other aircraft
- 168 blimps
- Six rubber plants (of its own and for the Defense Plant Corp.)
- Construction and operation of the world's largest powder-bagging plant for artillery explosives
- More than a million bullet-sealing fuel tanks for land vehicles and aircraft
- 200,000 rubber boats, life rafts, and pontoons
- 125,000 inflatable Mae West life jackets
- 4,880 barrage balloons
- Thousands of tons of tank tracks and bogey wheels

More than 26,000 Goodyear employees served in the armed forces of the United States and its Allies, and at home thousands of Akron women worked on the company's production lines. They came from nearly all life-styles and ranged in age from high schoolers to grandmothers.

Goodyear tire production, reduced to meet only military and essential civilian needs, exceeded 36 million from 1942 through 1945.

The company had a leading role in the all-out wartime effort to develop synthetic rubber — especially because of its pioneering work in this field — when natural rubber supplies from the Far East dwindled to a fraction of prewar days. By October 1945, 89 percent of the rubber used in U. S. industry was synthetic, compared with almost none at the start of 1942.

Gas masks were the first items Goodyear made expressly for war. Late in 1938, the Army had awarded the company a development contract for an improved mask, calling for the design of all necessary parts for a universal version — one that would snugly fit almost any face — plus the tooling required in large-scale production. Herman Morse, manager of industrial products development, headed this project team, which included Sam DuPree, a Georgia Tech graduate who had joined Goodyear in 1934 as a member of the production squadron.

A pilot line was set up in April 1939, and just before the Nazis invaded Poland, the Army asked for 200,000 complete masks a month. This department, which also supplied critical gas-mask parts to other manufacturers, was quickly expanded to 2,000 employees, more than half of them women.

Technology in light metal alloys, developed for use in the airships *Akron* and *Macon* and blimp control cabins, was responsible for the first military aircraft assignments. Glenn L. Martin Co., which produced B-26 *Marauder* all-metal, twin-engine bombers, asked Goodyear to design and build B-26 ailerons, flaps, and complete empennage sections, awarding a subcontract for this work on December 5, 1939.

## Goodyear Aircraft Corp.

On the same day, a new wholly owned subsidiary, the Goodyear Aircraft Corp. (GAC), was incorporated with 140 employees. By mid-1943, its payroll had increased to 32,000. The Goodyear-Zeppelin Corp., which had been formed in 1923 to build rigid airships, was dissolved on February 14, 1941.

The huge Goodyear Airdock, idle for years, provided plenty of room for the manufacture of airplane parts. Dr. Arnstein and his cadre of aeronautical and mechanical engineers, diverted to more mundane projects when the rigid airship industry died out, formed the backbone of this new venture. Even though Martin supplied basic specifications for loads and dimensions, the contract called for detailed design work by Goodyear's engineers.

GAC produced components that exceeded contract requirements for strength and quality at a cost 10 percent less than originally estimated, powerful ammunition for a sales force scrambling to get prewar subcontracts.

The U. S. economy accelerated under orders for war commodities in Europe, and Goodyear expanded production to keep pace. It began construction in late 1939 on a plant at Saint Marys, Ohio, to make industrial and automotive rubber products and Pliofilm. This facility produced its first molded goods on January 2, 1940, and Pliofilm output started in late March. It had 66 employees at the outset, but as business multiplied, an expansion to double capacity was quickly installed; the year ended with more than 700 on the payroll.

As the drums of war beat more insistently in Europe and Japan girded for military expansion in the Far East, Goodyear's shipments became increasingly military oriented. Sur-

prisingly, production of goods for peaceful purposes also expanded at furious rates.

Among the latter — and among the most unusual tires ever to come from any Goodyear factory — were six tires supplied to the Second Antarctic Expedition of Admiral Richard E. Byrd. These tires, the first of which was Goodyear's 300 millionth pneumatic tire, were 10 feet in outside diameter and almost 3 at the cross section.

The largest tires Goodyear had made, they were mounted on the expedition's snow cruiser, a vehicle 55 feet long, 12 high, and 15 wide, which was mobile headquarters for a crew of four in the Antarctic exploration. The company had produced prototypes of the Byrd tires in 1937 for Gulf Oil Corp. for use on large vehicles working marsh areas in search of oil.

Other unusual sales of the time that reflected continuing diversification were 125,000 square feet of rubber flooring in 25 colors for the steamship *America*, largest merchant vessel built in the United States; 236,000 feet (more than 44 miles) of single-jacket fire hose for use by the British in fighting air-raid fires; and 20 miles of conveyor belting to transport aggregate from the Sacramento River in Northern California to the Shasta Dam construction site 10 miles away.

When war erupted in Europe, about 18,000 of Goodyear's 46,000 employees worked outside the United States. Many who were citizens of countries involved in the fighting went into military service. Some members of U.S. families at the English factory were brought home, and the plant there was placed under supervision of the British government, concentrating on truck tires and industrial products.

The Swedish plant reduced operations by 50 percent when pleasure vehicles were ruled off the highways, but within two months was working at capacity on truck-tire production. Mechanical goods operations in Canada, England, Australia, Argentina, and the United States were soon at capacity and in back-order situations.

## Thomas, DeYoung Move Up

Just before the U.S. entry into war, two men who would lead the company through the next three decades moved upward in the management ranks. In 1939, Russell DeYoung was awarded a Sloan Fellowship at MIT where he received a master's degree in business administration; he was then made assistant to Litchfield and given the responsibility for supervision of Goodyear Aircraft. As war materials orders increased, he was put in charge of production at GAC.

On September 1, 1940, E. J. Thomas became Goodyear's eighth president. He was elected just a few months after his 41st birthday and the 24th anniversary of his employment. Litchfield continued as active chairman of the board, but assigned full responsibility to Thomas for executing policy and directing operations. Thomas concentrated his attention on the tire and rubber business, and, during the war years, Litchfield spent mornings at the Market Street headquarters and afternoons at GAC a few miles south.

*The Byrd snow cruiser. The largest tire Goodyear had ever made until that time outfitted the snow cruiser that bore Admiral Richard Byrd's expedition to the Antarctic in 1928.*

Two months earlier, Howard L. Hyde, a Cleveland lawyer, was elected general counsel and assistant secretary of the company, replacing Fred Wahl, who died in November 1939 after 15 years as head of the law department.

The rapid adaptation of Goodyear facilities and personnel to the production of diversified military items in the two years preceding Pearl Harbor was more than a tribute to Goodyear talent. It was a good example of the ingenuity and flexibility of all American industry.

GAC's convincing work as an airframe manufacturer was noticed by many major aircraft makers. In September 1940, Grumman Aircraft Co. contracted for complete empennage sections for the TBF-1 *Avenger* torpedo bomber. In October, Curtiss-Wright Co. asked for stabilizers for its P-40 *Warhawk* fighter plane, and the Navy ordered four K-type patrol blimps and two L-type training blimps.

In November, Martin awarded a subcontract for various control surfaces of its PBM-3D *Mariner* flying boat, and Consolidated Aircraft Co. subcontracted for tail sections, ailerons, flaps, outer wing sections, and floats for its PB2Y-3R *Coronado* long-range patrol flying boat. By the end of the year, the vast floor space of the Airdock was totally committed to production, and additional space soon would be required.

## New Plants, More Contracts

The company's speed and efficiency in adapting to the defense effort had a positive effect on military procurement agencies. Litchfield and Thomas made it known that Goodyear had the will, the know-how, and the skilled manpower ready to undertake almost any assignment.

One other major prewar military recognition of Goodyear's expertise was an Army contract to lay out and operate a $35 million bag-loading facility on the Ohio River near Charlestown, Indiana. Known as the Hoosier Ordnance Plant, it provided bags of smokeless powder supplied by the Du Pont Co. from an adjoining factory. The finished product was propellant charges that ranged from 3 to 800 pounds for 60 different types of artillery.

The Goodyear Engineering Corp. was set up in late 1940 as a subsidiary to operate the Hoosier plant. Harry Hillman, who had joined Goodyear in 1934 as assistant treasurer, became vice-president in charge, and more than 40 engineers, production supervisors, and personnel specialists were transferred to Charlestown as a nucleus for manning the project.

Expansions and new defense contracts became commonplace in 1941 as the national defense program gained momentum. In January, Goodyear announced it would join Martin, Chrysler Corp., and Hudson Motors in a joint program to produce 100 B-26 *Marauder* medium bombers on a monthly basis. GAC would build the wings and engine nacelles. This would require 7,000 employees, more than 10 times the GAC payroll at the end of 1940. Its Akron facilities were expanded with two plants (designated B and C) adjacent to the Airdock for making airplane wheels, brakes, and parts.

The next GAC expansion was in response to an appeal from Lt. Gen. William S. Knudsen, who had resigned the presidency of General Motors to become production chief for the Army. Knudsen wanted even more aircraft parts from Goodyear. Litchfield, backed by Thomas, suggested a factory site at Litchfield Park, and Knudsen, skeptical about the labor supply there, finally agreed to it. The plant was financed by the government and leased to Goodyear. Construction was started in midsummer 1941 in the Goodyear Farms section of Litchfield Park, and production began in February 1942 when payroll reached 500. Through the war years, payroll grew rapidly, and by early 1945 the facility had 7,500 workers.

Manufacturing space continued to be a

*Goodyear workers, including a few out of the thousands of Akron women hired to relieve the wartime labor shortage, engaged in B-26 production. The company's first military aircraft assignments were through a subcontract to produce parts for B-26 Marauder twin-engine bombers.*

problem for the corporation. So in May 1940, the gymnasium and entire top floor of Goodyear Hall were converted to balloon manufacturing, along with a long-dormant textile plant in New Bedford, Massachusetts, which had been acquired from Rotch Fabric Mills in 1924.

Proud of its prewar production achievements, Goodyear did not hesitate to publicize them. In 1940, a 20-minute sound movie on the company's defense work, *Goodyear Shoulders Arms,* was produced with the cooperation of Pathe News. It was turned out mainly to accompany an exhibit of products manufactured for the armed forces and displayed at dealer conferences, employee meetings at all major Goodyear installations, and industrial trade fairs.

Following the film's premiere in Chicago for 250 dealers and 91 news media representatives on January 15, 1941, a special presentation was made in Washington, D.C. Guests included Adm. Chester Nimitz, chief of the Bureau of Navigation and soon to be commander in chief of the U. S. Pacific Fleet; Paul McNutt, administrator of the Federal Securities Agency; William McReynolds, administrative assistant to President Roosevelt; Donald Nelson, director of purchases for the Office of Production Management; and numerous senators and congressmen.

By June 1941, GAC had 1,320 employees, nearly all in defense work. On June 4, the first Aircraft Edition of the *Wingfoot Clan* was published. The five-column, page-one banner headline read, "America First In The Air Is Our Slogan."

"We are dedicated to the task of making AMERICA FIRST IN THE AIR," Litchfield wrote in the editorial. "To that end we must work together — determinedly, efficiently, and harmoniously . . . Unless and until America is the most powerful nation in the air, our safety, our freedom, and our standard of living will not again be what they have been in the past."

When personnel reported for work at all Goodyear factories and district offices in the United States on Monday, July 28, they were greeted by new metal three-color reproductions of the time-honored slogan originated in 1915: "Protect Our Good Name." The signs, hung over the weekend, were highly visible in every domestic office and work place.

As 1941 neared its end and "the day that will live in infamy," Goodyear's defense production had reached proportions almost beyond imagination a few years earlier. But the work for war had only begun.

## The Hungry War Machine

Three days after December 7, when Japanese bombers attacked Pearl Harbor, crippled the U.S. Navy fleet, and killed 2,403 Americans, Goodyear stated its commitment to war in a page-one Litchfield editorial of the *Clan.*

> The hour of war has struck.
>
> Our people have been attacked by a foreign foe.
>
> They must be given blow for blow — and more.
>
> I do not say we must unite — I say we have united.
>
> Internal differences are automatically resolved.
>
> Capital, labor, and management are as one, shoulder to shoulder in common and paramount determination — to wipe those sinister forces from the face of the earth.
>
> Every resource and facility owned by Goodyear, every bit of our talent and manpower, are the President's to command.
>
> This is offered without reservation or qualification.
>
> As individuals and as a corporation, we have worked hard during the past months to be ready for this situation. Now we shall double and redouble our efforts.

Before year's end, GAC was selected as subcontractor to produce wings and tail sections for Northrop Aircraft's P-61 night fighter, the *Black Widow*. On December 31, DeYoung announced that steel was going up for a large addition to Plant B, even though the original building was incomplete.

"We are doing everything we can to increase production rapidly," he said. "We are now operating three eight-hour shifts, seven days a week, and New Year's Day will be no exception.

"We are training men as fast as possible and adding personnel to all operations just as soon as trainees and skilled operators are available. The payroll, now in excess of 4,000 employees, will be increased to 10,000 or more."

But 10,000 would prove less than enough as the hungry war machine demanded more and more material and GAC's sales force under sales manager Thomas A. Knowles worked hard for a fair share of the booming aviation products business. The subsidiary made administrative realignments to direct the increasingly complicated operations. Harry E. Blythe, a 27-year veteran, was made vice-president and general manager; DeYoung was named vice-president, production; Dr. Arnstein, vice-president, engineering.

In the company's U.S. tire factories, total employment had grown from 11,550 in 1940 to 15,817 by the close of 1941. The defense buildup probably was reflected in the 1941 annual report to shareholders: sales of $330.6 million were up 52 percent over 1940.

## A Grim Christmas

The Christmas season of 1941 was one of the most somber in U.S. history, perhaps the least cheerful since the Civil War. By New Year's Day, Japanese troops occupied Guam, Hong Kong, Borneo, Wake Island, and the Philippines. Their bombers regularly attacked Britain's island fortress of Singapore, which would fall in six weeks. German U-boats torpedoed American merchant ships off the U.S. East Coast, and their troops were at the gates of Stalingrad, preparing for a final attack on Moscow. Germany's General Rommel threatened Cairo.

Obviously, 1942 started on a grim and worrisome note, but industry was confident. Faced with the greatest challenge in the nation's 166 years, it would soon be ready to meet that challenge.

Most of Goodyear's foreign subsidiaries already had gone to war. The English plant at Wolverhampton, in the heart of the industrial Midlands and an inviting target for Germany's Luftwaffe air squadrons, operated at capacity on war production orders. Factory employees worked 20 days in a row, then got one day off. Walter Hazlett, the managing director, commuted by bus between home and plant, slept in a dugout in his backyard, and drove an ambulance 2 nights a week.

The plant in Indonesia was earmarked for defense needs of the Netherlands East Indies. The Australian factory concentrated on tires for military vehicles. Factories in Argentina and Brazil reduced production because of diminished export trade with Europe. The Canadian factories produced almost totally for the war effort; the general manager and treasurer of Goodyear-Canada, R.G. Berkinshaw, had moved to a government position as director-general of the priorities board of the Department of Munitions and Supply.

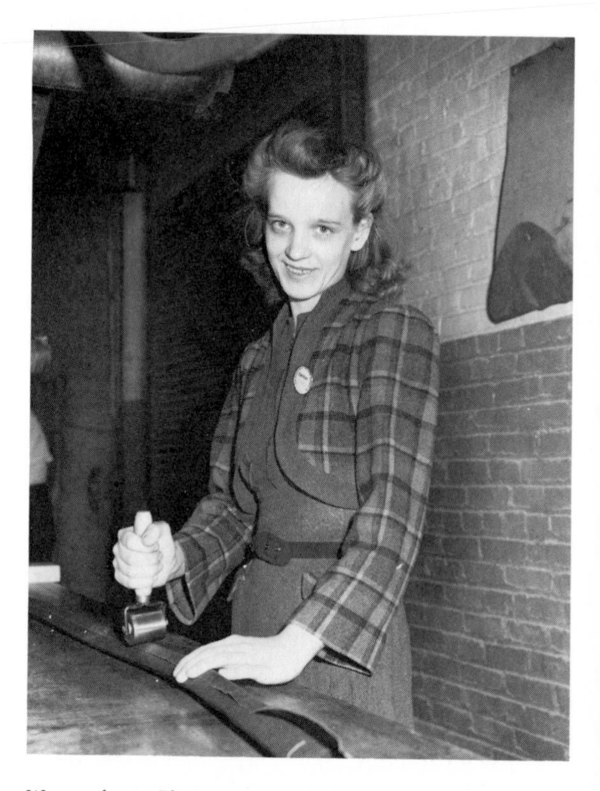

*War worker in Plant II. Women played unaccustomed roles when thousands were asked to fill production jobs vacated by Goodyearites called to military service. Military contracts created a need for tens of thousands of new workers while over 26,000 Goodyear employees served in the armed forces.*

In the United States, Personnel Director Fred Climer was appointed to membership of the federal War Production Board and assigned to joint labor-management committee programs. Dr. Ray Dinsmore, Goodyear's top scientist, was named to the staff of the National Rubber Administrator to help coordinate a national effort to develop synthetic rubber.

## The Synthetic Miracle

The tremendous increase in synthetic rubber production in the United States ranks as one of the great achievements of industry. From a mere trickle of this new product in 1942, the nation turned out as much synthetic rubber in 1946 as the total of natural rubber it consumed in 1941.

At the beginning of the war, the United States had a stockpile of about a year's supply of natural rubber and no plants that could even approach making synthetic on a large-production basis. But industry did have the savvy to produce it — in the laboratory or in small pilot plants, such as Goodyear's. In 1940, Litchfield had asked the directors for $250,000, a figure he later boosted to $400,000, to raise that plant's capacity for Chemigum production to 2,000 tons a year, even though that still fell short of commercial production range. When the call came, though, Goodyear was ready with the technology to manufacture the much-needed replacement for natural rubber.

In 1940, the U.S. Government Rubber Reserve, an agency set up primarily to accumulate a stockpile of natural rubber, asked each of the Big Four companies to build a synthetic rubber plant with a 10,000-ton annual capacity. In May 1941, the Office of Production Management — another federal group — recommended that each company raise this capacity to 40,000 tons. Immediately after Pearl Harbor, when shipping routes from the Far East had been jeopardized, the Rubber Reserve directed that the 40,000-ton program be increased to 400,000 tons for each company.

The Rubber Reserve consulted with the rubber industry, and together they decided that the most practical synthetic rubber for the war effort was a general purpose compound containing Styrene and Butadiene. Called GRS, for Government Rubber, the synthetic's production bounced from 3,271 long tons in 1942 to 719,404 in 1945.

In addition to expanding its own synthetic rubber plant in Akron, Goodyear supervised the construction of three synthetic rubber plants for the government — at Houston and Baytown, Texas, and Torrance, California. Initial annual capacities would be, respectively, 60,000, 30,000, and 60,000 long tons. Goodyear was commissioned to operate the Houston and Torrance factories; Baytown was assigned to General Tire. Further, the government ordered all synthetic rubber produced in the United States into a national stockpile for allocation by the Rubber Reserve.

## Shortage of Passenger Tires

Late in 1943, the company produced the first synthetic tire made entirely of government-produced materials. The rubber was made at the Akron Chemigum plant, the most advanced and highest-producing facility for synthetic rubber. All Akron rubber companies were then well into synthetic rubber manufacturing. Firestone operated two plants in Akron and one at Port Neches, Louisiana; Goodrich was in operation at Port Neches, Louisville, Kentucky, and Borger, Texas; General and three smaller companies operated a plant at Baton Rouge, Louisiana.

That May, Goodyear announced the 'S-3, a synthetic rubber tire for passenger cars. It had been authorized in two popular sizes by the War Production Board and was a big improvement over the War Tire, made entirely of reclaimed rubber. The acute need for passenger tires for essential transportation was described in a progress report that month by William M. Jeffers, rubber director for the war board.

> By 1944, the country will have gone two years with less than one quarter of the normal replacement of tires and with no new cars. This accumulated deficit indicates that new tires must be provided to keep the country moving. Surveys show that 30 thousand is a probable minimum replacement program that the country can get by with, even by general recapping, by maintenance of present driving speeds, and by keeping conservation measures.

Jeffers called for fabrication of at least five million new tires in 1943 for use by "essential drivers."

## 168 Blimps; 4,008 *Corsairs*

In February 1943, a Navy fighter plane took off from Akron Airport on its initial test flight. It was the first of the 4,008 Vought *Corsair* FG-1 fighter planes built by Goodyear during the war years, and it took to the air just a year after the company had received a go-ahead for its first and only production of a complete plane.

The *Corsair,* designed and first built by Chance Vought, had a 2,000-h.p. engine, range of 1,000 miles, ceiling of 35,000 feet, top speed of close to 400 mph, and low landing speed of 85 mph. It carried six 50-caliber machine guns in the wings, which the pilot could fire singly or in salvo, and attachments for two 1,000-pound bombs. It was a carrier-based aircraft and was just less than 17 feet tall with its wings folded.

To build this plane in required Navy volume, Goodyear constructed a new plant 1,450 feet long and 450 feet wide, a flight hangar 800 x 300 feet to house 70 planes at once, a concrete apron, and a gun range. Taxiways were built to connect this plant with Akron's Municipal Airport, and a new work force of 10,000 was hired.

For a company that had never built a complete plane, this was indeed a challenging assignment, but Goodyear and its people jumped into the job with enthusiasm. Its record in producing airplane parts before the war obviously had influenced the Navy in its selection.

The contract was signed in February 1942 and for reasons of security received no publicity. All Goodyear publications remained silent about the *Corsair,* even when plant construction had progressed to where tooling was under way and A. C. Michaels and W. C. Potoff had been supervising the project for months.

Hugh Allen, a veteran newsman and author who had been in the company's public relations department since 1919, described the extensive subcontracting necessary for the *Corsair* in his 1947 book, *Goodyear Aircraft.*

> To turn out its 4,000th *Corsair* two-and-a-half years after it turned out its first one, Goodyear Aircraft subcontracted 40 percent of the job. Some of the subcontractors were worldwide corporations, others were one-man machine shops. They were scattered all the way from New England to California.
>
> In looking for companies that might supply parts or equipment . . . Aircraft turned first to those industries whose normal business had been upset by war . . . It went to manufacturers of stoves, ambulances, linoleum, metal furniture, tin cans, glassware, caskets, wrapping paper, fishing tackle, hosiery, brassware.
>
> Makers of thermometers changed over to build valves; washing machine companies built hydraulic cylinders and actuating control rods; an outdoor advertising company made wheel subassemblies; an Akron department store, the M. O'Neil Company, made wing coverings, map covers, and headrests.
>
> Goodyear set up machine shops of its own, but sent everything outside that it could . . . A manufacturer of golf clubs at Newark, Ohio . . . leased his plant to the company for machining of forgings and castings . . . Garages and car dealers changed their service departments into factories. Mechanically inclined individuals and retired machinists set up shops in their basements. Farm wives gathered at each other's homes, sorting rivets on piecework.

Recruiting a work force for the *Corsair* project and other war work at GAC when so many men were in uniform was just as difficult as lining up subcontractors. Goodyear offered housing, training, and transportation benefits,

and GAC's payroll climbed swiftly from 3,500 at the time of Pearl Harbor to 32,000 just 18 months later.

Women made up more than 50 percent of the total work force. Paul Scheiderer, later to become personnel manager at GAC, recalled how he recruited young women in West Virginia. "I asked two questions," he said. "Can you harness a team of mules and milk a cow? If the answers to both were 'yes,' I hired them, figuring they had the dexterity to work in the plant." Retired machinists, insurance salesmen, politicians, war brides, housewives, and high schoolers operated drills and rivet guns. Midgets and dwarfs proved useful in airplane wings and pontoons of flying boats, holding rivets in place as workers on the other side drove them home. About 20 blind people at GAC sorted small parts, nails, and screws and handled other tasks that could be accomplished entirely by feel; their seeing-eye dogs sat patiently beside them amid the factory din.

Facilities outside Akron also were cogs in the war production team. The plant at Jackson,

*Right, midgets and dwarfs were particularly suited for work in production of aircraft parts. Below, women war workers finish up production of Goodyear's 1,000th Navy Corsair, 1944. The company built 4,008 Vought Corsair FGI fighter planes between February 1943 and August '45, subcontracting 40 percent of parts production to companies whose business had been hard-hit by the war. Opposite, GACers build a Vought Sikorsky fighter plane.*

Michigan, considered the most modern tire factory in the world, was cleared of all tire-making equipment to make room for gun-manufacturing machinery. Its last tire was cured on April 27, 1942, and first antitank cannon turned out on November 5.

The Kelly-Springfield factory at Cumberland made high-explosive artillery shells and armor-piercing machine-gun bullets. At the peak of production, more than one million bullets were produced each day, and of the 5,000 employees there, half were women.

At the new Saint Marys plant, automotive molded goods and Pliofilm were replaced by rubber tank treads, rubber gaskets for aircraft and ground vehicles, waterproofing cloth for Army tents, and packaging film for storing military items. Employment there doubled to more than 1,600 at the production peak in 1944.

The Gadsden plant made truck and tractor tires and tubes for all kinds of military and essential agricultural vehicles. Employment increased from 1,300 in 1941 to more than 3,500 in 1944.

The California installation in Los Angeles produced tires for military and essential civilian needs and a diverse range of military items. These included bullet-sealing fuel tanks, barrage balloons, gas masks, rubber boats, life preservers, and inflatable pontoons. Prewar employment averaged 1,300; it rose to nearly 5,000 by 1944.

Four textile mills, in Rockmart, Cartersville, and Cedartown, Georgia, and Decatur, Alabama, concentrated on cotton and rayon for tires, but they also supplied special fabrics for duffel bags, cartridge belts, tents, and field packs.

## Blimps on Coastal Patrol

When the United States was brought into World War II, its defense against submarine attacks on shipping and shore installations was dangerously weak. American freighters were being sunk almost within sight of the Atlantic Coast. A Japanese submarine attacked and sank the S.S. *Medio*, a cargo vessel, off Eureka, California, two weeks after Pearl Harbor; enemy subs shelled a coastal oil field farther south, near Santa Barbara, on February 23, 1942.

The Navy's fleet of blimps totaled only 10 units then, and all were based at Lakehurst, New Jersey. Goodyear operated 5 L-type blimps that were quickly drafted. The *Resolute*, operating in Los Angeles, was armed and in

*Above, rubber boat production. Goodyear built 200,000 rubber boats, life rafts, and pontoons during the war years. Opposite, hundreds of Goodyear employees participated in a "Land of the Free" pageant highlighting a rally staged by a labor-management committee as part of a production drive.*

service even before completing the legal technicalities of swearing in the crew and commissioning. This made the crew members temporary pirates aboard a privateer, but international protocol was not of much concern. The Navy needed blimps, lots of them, to help patrol and protect the coastal waters.

One of the first Goodyear blimps in Navy service — the *Ranger*, designated L-8 by the Navy — was involved in a mystery that never has been solved. Early on the morning of August 16, 1942, the airship took off from its base at Treasure Island in San Francisco Bay for an offshore antisubmarine patrol mission. It carried a crew of two, Lt. (jg) Ernest D. Cody and Ens. Charles E. Adams. Two depth bombs were racked beneath the control cabin.

Several hours later, residents of Daly City, a San Francisco coastal suburb, were astonished to see the L-8 gracefully descend to a soft and perfect landing in the middle of a residential street.

Naval personnel rushed to the scene. If volunteer firemen had not ripped open the envelope, the *Ranger* would have remained airworthy. There was fuel in the tanks, power in the batteries, the emergency life raft and parachutes were on board, and the radio was operating.

Everything seemed in order except the cabin door, which had been secured in an open position. One depth bomb was missing.

So was the two-man crew.

The Navy never developed a satisfactory explanation about what happened to the two officers, who had been wearing regulation life jackets when they earlier left the base, and the *Ranger*'s mystery remains totally unsolved.

In early 1942, Goodyear's blimp-building program became just as hectic as its other big war jobs. It was the only builder of blimps. The Navy ordered 23 immediately after Pearl Harbor, and Congress soon upped authorization to 200. The new production goal was an airship every two days.

Work started in March to double the size of the hangar at Wingfoot Lake; the company's hangars at Chicago and New York were dismantled and reassembled, one between plants A and B of Goodyear Aircraft and the other beside the Wingfoot Lake hangar. Two additional floors of Plant I in Akron were converted to envelope construction, and the Goodyear Hall gymnasium underwent another face-lifting, this time for making blimp control surfaces.

In June, the Navy increased its orders for K-model submarine patrol ships to 48 and asked for one L-model and 10 G-model ships for use in an accelerated training program to qualify 1,500 blimp pilots.

By September, nearly all the company's blimp pilots, who held commissions in the Naval Reserve, were on active duty to help train naval pilots or take key positions in the rapidly expanding lighter-than-air forces. More than 2,800 workers were building blimps at GAC, the tire plants, and Wingfoot Lake by the end of 1942.

As the blimps came off Akron production lines in increasing numbers, U-boat commanders became wary of seeking targets within sight of American shores. The new M blimps, 50 percent larger than the K model and designed for long-range patrol, were especially effective against the undersea marauders.

Navy airships compiled an astonishing record during the war, especially as aerial escorts for convoys. In U.S. Atlantic and Gulf Coast waters and in the coastal waters of the Caribbean, Eastern Central America, and Brazil, 532 Allied vessels in all were sunk, but not one by enemy submarines while under airship escort. Altogether, Navy airships escorted 89,000 sea vessels during the war without loss of even one.

Only a single blimp went down because of enemy action. On July 18, 1943, K-74 detected and engaged an enemy sub in the Caribbean. The U-boat's gunfire, momentarily silenced by the airship's machine gun, slowly brought down the K-74 when her bombs failed to release. The big airship floated for hours, and all but one of her crew were rescued the next day.

Major blimp production in Akron was suspended in April 1944 as the number of airships in service was considered more than adequate to cope with the greatly diminished U-boat fleet. In all, Goodyear had supplied 132 K-, 22 L-, 8 G-, and 4 M-model ships for coastal defense during World War II.

*Formation of Navy "L"-class training ships, 1944. Goodyear, the only U.S. builder of blimps, was pressed into service to meet the urgent need for blimps to patrol and protect coastal waters. During World War II, the company supplied 132 Navy airships and trained Navy airship pilots. As aerial escorts, the blimps greatly reduced the threat of enemy U-boats.*

Despite the near shutdown of lighter-than-air craft production, there was no letup at Goodyear Aircraft. The nation's need for more and better airplanes continued, and GAC was in the thick of competition to meet those needs. Undoubtedly, the company's blimp-building record had turned into a plus for Tommy Knowles, elevated to GAC sales vice-president in 1944, and his wide-ranging sales force.

## The Phantom Fleet

One of Goodyear's most unusual and romantic wartime accomplishments was committed to absolute secrecy until the conflict had ended. It was the phantom fleet for the Allies "Grand Ruse," which helped confuse Nazi air reconnaissance before the D-Day invasion of France.

For several weeks preceding June 6, 1944, amphibious invasion craft, P T boats, tanks, combat vehicles, and heavy artillery were seen by German intelligence sources at several different English Channel and North Sea ports and beaches. The confusing element was that as those sites seemed loaded with all this equipment for a day — or a few days — they might suddenly appear abandoned while other ports apparently swarmed with invasion equipment.

This great deception, which may have influenced the erratic disposition of German coastal defenses, was made possible by inflatable fabric replicas manufactured by Goodyear at its Woonsocket, Rhode Island, plant and by a few other rubber companies. The same Goodyear balloon experts who had produced giant comic figure balloons for New York City's traditional Thanksgiving Day parade had responded to the military's urgent request for reproductions of heavy equipment and about a dozen different basic craft.

Built on a one-to-one scale and with all appropriate markings, the reproductions were rushed to England where they were inflated to form fleets of invasion craft — all dummies and all armed with weapons that could not fire a shot. They appeared and disappeared, often overnight, in great numbers along the coastlines, and shifting them from one port to another was easy. Deflated after dark, they were folded into compact packages, loaded onto trucks, and hauled to new destinations where they were reinflated before daylight.

The project to produce part of the phantom fleet was labeled high priority and top security. The 600 Goodyearites working on the inflatable replicas had no idea of their end use, and the military took delivery of the finished products without giving a hint of their ultimate purpose.

Throughout the war, many Goodyear installations in the U. S. were in a state of flux, adjusting to meet military requirements. New contracts and frequent revisions of specifications for components already in mass production called for almost constant alterations of plant layouts and tooling.

When the Curtiss P-40 *Warhawk* project began phaseout in late 1942, space was cleared for equipment to produce parts for the Curtiss SB2C scout bomber. Before that assembly line could get into full production, though, the entire job was transferred to a new Curtiss plant in St. Louis, Missouri, to provide more space for the expanding blimp-manufacturing program. Early in 1943, even more room for blimp building was created by moving the P-61 *Black Widow* assembly line from Plant B to Plant C.

A much more complex shifting of assignments occurred in July 1943 to phase out the Martin B-26 wing project and install machinery for the production of immense new bomb-bay sections and empennages. The aircraft that would use them was not disclosed at first, but it was soon revealed to be the Boeing B-29 *Superfortress*, the bomber that went to war in the Far East and eventually dropped atom bombs on the Japanese cities of Hiroshima and Nagasaki in August 1945.

While Goodyearites worked on parts for a host of military aircraft and finally the B-29, Charles J. Pilliod, Jr., of neighboring Cuyahoga Falls, joined the company as a trainee in 1941, entered the Army Air Corps the following year, and eventually went into training as a pilot of the new *Superfortress* bombers. In 1974, he would become Goodyear's fifth chairman of the board.

*Above, war workers put the finishing touches on dummy trucks. Below, replica planes in position. Inflatable rubber reproductions of military tanks, boats, planes, and artillery, moved to ever-changing positions under cover of night, confused and deceived the Nazis for several weeks before D-Day, 1944.*

*From left to right: Edwin J. Thomas, president, and Paul W. Litchfield, chairman, Goodyear Tire & Rubber Company; John Thomas, chairman, Firestone Tire & Rubber Company; William Jeffers, Rubber Director; John Collyer, president, BF Goodrich Co.; Frank A. Seiberling, chairman, Seiberling Tire & Rubber Co.; Harvey Firestone, president, Firestone. 1942.*

## Tire Leader in War

Although the aircraft subsidiary had the most glamorous Goodyear role in the war effort, particularly in the early stages, the company's basic business — tires — made a contribution of exceptional importance. Goodyear produced more tires in World War II than any other company. At the outset, tire production was reduced drastically in the quick conversion to the manufacture of armaments. The 1941 output of 17 million was cut back to about 5 million in 1942, then boosted to 6.75 million the following year.

By mid-1943, tire industry production could not supply both the mechanized military complex and domestic requirements, and motor transportation was increasingly essential to both. Some Goodyear plants were therefore being reconverted from military production by the second half of the year.

The big Jackson factory that had concentrated on antitank guns and fuel tanks now turned out tires as well. By December, tires were its only product. The Kelly-Springfield plant continued to make high-explosive shells, but produced its last machine-gun bullets in September when it resumed tire making.

The Gadsden factory was nearly doubled in size for increased tire output, and the Los Angeles plant went back to full tire production.

In August 1943, President Thomas addressed supervisory personnel gathered at the Goodyear Theater in Akron.

> There's lots of war to fight yet. The loyal cooperation and utmost effort of every Goodyear man and woman is necessary. Thirty million synthetic passenger car tires and six and one half million synthetic truck tires are needed for essential driving alone . . .
>
> We are now called upon to produce a normal peacetime volume of civilian replacement tires with synthetic rubber, of which we have much more to learn. We knew the answers when we used crude rubber, because we had 40 years in which to study that substance.

Learning went apace, and so did tire manufacturing, but by mid-1944 the large-scale industry reconversions were insufficient to meet demands. In December 1944, Gen. Brehon Somervell, the Army's chief of supply, told an emergency conclave of top executives and labor leaders of the rubber industry that the Army would be 472,000 tires short for the first quarter of 1945.

The industry responded by pledging a 120-day drive to maximize machinery and manpower. By mid-February of 1945, military tire output at Goodyear's Akron plants was at the highest level of the war period, and more than 9,000 manufacturing employees had worked every day since the start of the four-month push in December.

Years later, Thomas recalled the emergency.

> Tire output at first was cut back to a level deemed adequate only for the military and essential civilian requirements. The initial step was to convert many of our plants from normal production to manufacturing of war items.
>
> But ultimately some of them had to convert back to making tires for the war effort. This amounted to a double conversion during the war period. Our attitude was, "Tell us, Government, what you need most to win the war, and that's what we will do."
>
> We not only had to convert some plants back to tire making, but build additional tire plants to keep the war machine going.

Backed by the federal Reconstruction Finance Corp., Goodyear started a new tire plant at Topeka, Kansas, in the fall of 1944. It produced large-sized tires for combat vehicles and rear tires for farm tractors. The company purchased this plant from the government in September 1945.

## Union Unrest

Goodyear Aircraft went through the war without any production stoppages because of labor disputes, but the Akron rubber plants had two strikes. The first expressed dissatisfaction with wage guidelines established by the War Labor Board and lasted five days.

The second involved a wide range of grievances and alleged contract violations and started on June 17, 1945. It closed Goodyear's Akron rubber plants for 18 days; inspired criticism by the government, news media, the public, and some other unions; evoked disavowal by the international union; and in the end forced government to take over. Government stepped in, however, only after the Secretary of War, the Secretary of the Navy, and the governor of Ohio urged a return to work and the Selective Service Board threatened to draft any strikers exempted as essential war workers.

In spite of popular opposition, the union membership voted 8,561 to 3,039 to strike, and the plant shut down from June 17 to July 5. Legally unable to force the Akron local to accept the War Labor Board's injunction to return to work, the government had to move in, and President Harry S Truman ordered the Navy to assume control of the situation. The strikers returned to work.

Goodyear's president and the naval officer in charge reached a first-day agreement on running the plant. E. J. Thomas pointed out that the takeover was a new company experience, and a distasteful one.

"Now, Captain Clark," he said, "as of this minute who is running the plant? Are you running it, or am I running it? I think we ought to determine that first."

"Well," replied the captain, "you are, but technically it is in possession of the Navy. Obviously we can't run this business, and we want you to run it."

"Do you want to take over my office, or shall I find one for you?" Thomas asked.

"All we will need," said the captain, "is a place where my personnel and I can operate."

Thomas found a place for the naval men and later recalled: "We went right on from that point and ran the plant and the business. The Navy representatives were a tempering force and did some good work in straightening out some people who needed straightening out. But the heat went out of the whole thing pretty fast, and our normal operations resumed."

The Navy returned the Akron plants to Goodyear on August 31, 1945, after less than two months of military control.

## Parts for the Superfortress

When May 7, 1945, signaled the end of the fighting in Europe, nearly all of GAC's production went to war in the Pacific. Experience gained in the hectic early war years and improved assembly-line techniques permitted a work-force reduction to about 18,500, releasing thousands of employees for other war jobs and military service. The size and scope of GAC's effort in the conflict's closing days was indicated by the monthly production rates attained for major projects just before the final day of the war, August 15, 1945. Here are some examples as they appear in company records.

- 55 bomb-bay and empennage sections for the B-29 *Superfortress*
- 60 outer wings for Consolidated Aircraft's B-24 *Liberator* bomber
- 52 sets of wing sections and empennages for the P-61 *Black Widow* fighter
- 105 sections for the PV-2 *Ventura*
- 200 sets of parts for the Grumman G6F *Hellcat*
- 27 sets of control surfaces for the Martin PBM *Mariner*

In June, 225 FG-1 *Corsairs* were delivered just before a tooling changeover to make the advanced F2G *Corsair.*

Goodyear Aircraft's sales volume during the last two war years equaled that of the company's tire and rubber operations. Goodyear's wartime contracts for aircraft represented more than 45 million pounds of airframes.

The company's technological capability had increased tremendously during the war. Much of this progress had been initiated in the Goodyear Research Laboratory, which was built in Akron during the early war years and dedicated in June 1943.

This $1.3 million building housed 250 chemists, physicists, engineers, and other technicians working to find new ways of increasing and improving wartime production. A more lasting reason — the peacetime future — was expressed in the legend at the building's entrance: "The Best Is Yet to Come." In postwar years, the Research Laboratory became the major wellspring of technological innovation that would help maintain Goodyear's leadership in the international tire and rubber industry for many, many years to come.

## A Difficult Transition

As the world stepped tentatively into a new era in late 1945, Goodyear began its own postwar transition, a difficult and traumatic process. The company had to transform itself practically overnight. It had to shed the role of large aviation manufacturer and munitions maker and become once again a manufacturer of tires and rubber products. Further, it had to offset the almost immediate cancellation of $432.4 million of government contracts while girding for competition in growing markets.

Because Goodyear Aircraft operations had been directed almost totally to the war effort for nearly five years, it had to release thousands of employees when the war ended. The payroll dropped from 28,903 in January 1945 to 4,622 by year's end and to about 2,000 by mid-1946.

A high percentage of these workers neither were trained for nor wanted jobs in tire manufacturing, which was heading for record production. Some returned to their prewar occupations, and many women resumed their lives as homemakers. The need for rubber workers, however, was so pressing that recruitment once again extended into neighboring states as the Akron labor pool was soon exhausted.

Despite these difficulties, plus union demands for a reinstatement of six-hour work shifts in Akron, tire production was helped by the return of military veterans, and it soon rose rapidly. Goodyear's output increased by 25 million tires in 1946 and again in 1947 as the auto industry responded to the sudden demand for new cars and owners holding onto their old cars rushed to replace old tires.

Emerging from its wartime experiences and nearing its 50th anniversary, the company had changed its outlook and basic policies little from the early days. The entire corporation was still people oriented, and its people were proud to be Goodyearites. A visitor to any Goodyear operation could not help noticing "Protect Our Good Name" signs, constant reminders that the individual must do his part in the progress of the corporation.

P. W. Litchfield and E. J. Thomas were good assurance that the Goodyear family feeling remained alive. Firm and exacting managers, they were nevertheless nonauthoritarian and maintained old friendships throughout the company despite disparities in rank.

Thomas, an ebullient man with great personal charm, was known as Eddie to hosts of factory and office workers throughout the Goodyear world. He and Litchfield attended most company functions in Akron — including many intramural athletic events — faithfully participated in employee service anniversaries, and kept a close watch on potential managers.

Three of Litchfield's fellow pioneers, C. W. Seiberling, Clara Bingham, and Ed Hippensteal, died in 1946. All had joined Goodyear in its first year and were members of the Old Guard.

C. W., who had helped his brother build the Goodyear groundwork and was a friend to hundreds of Goodyearites in his 23 years in management, had continued as an honorary member of the Old Guard after leaving the company in the 1921 reorganization. Clara had hired in as a stenographer and for many years was secretary to the Seiberling brothers. Ed Hippensteal was Goodyear's first employee, hired before incorporation to help clean up the old strawboard factory.

In 1947, Cliff Slusser, who had been vice-president in charge of production for 21 years, moved aside for less arduous duties as vice-president and general manager of the textile and coal mine subsidiaries. He was succeeded by 37-year-old Russell DeYoung, who had made an excellent record as production vice-president at GAC.

Goodyear enjoyed a sellers' market in the three years following the war and established new records for peacetime sales: $616.5 million in 1946, $670.7 million in 1947, and $704.8 million in 1948. The old record of $330.5 million had been established in 1941 when defense business was reflected in the total.

The management team, aware that the sales tide would turn — perhaps even into a depression — concentrated on the intense competition ahead, cultivating the goodwill and respect of distributors and the public.

Typical of efforts to improve dealer relations and strengthen outlets was the Visual Merchandising Laboratory, described by Sales Vice-President R. S. Wilson as an "epoch-making step in the preservation of free enterprise." It provided expert merchandising instruction and architectural guidance for dealers of all sizes and categories.

The lab included full-size replicas of small, medium, and large dealer establishments, completely stocked and equipped with up-to-date fixtures and merchandise displays. Goodyear invited dealers to the laboratory, which had been established in Goodyear Hall, to see and learn how their businesses could be upgraded, modernized, and given more sales appeal. This training program became an integral part of the company's dealer development curriculum.

## Blimps Rally Round

Although postwar sales rode along at record highs, Goodyear resumed its aggressive promotion and marketing soon after the peace treaties had been signed. The L-model blimps *Ranger, Enterprise,* and *Volunteer* were bought back from the Navy in 1946 for advertising and promotion campaigns.

*Puritan,* a K-type airship and larger than the L models, was purchased from the War Assets Administration and renovated to display the largest aerial signs yet flown, 190 feet long with letters 18 feet tall. In May 1947, *Mayflower,* an L-type ship, also was returned and became Goodyear's fifth airship in service.

The K ships were soon withdrawn, though. The smaller L models were more flexible, required smaller crews, provided more operating economy, and made ideal advertising and promotion vehicles.

In addition to reviving the blimps as promotional tools, the company initiated new advertising fields. It first contracted with the National Broadcasting Co. to sponsor telecasts of nine major college football games, Goodyear's first venture into commercial television. The opening game on the telecast schedule was between Army and Cornell on October 5, 1946, over stations WNBT, New York City, and W3XWT, Washington, D.C.

## Public Service Programs

Next January 26, the company introduced its first public service radio program. Entitled "The Greatest Story Ever Told," it was a series of 30-minute narratives developed from the teachings of Christ in the New Testament. The program was broadcast on Sundays from coast to coast until 1956 over the American Broadcasting Co. network. It carried no commercial announcements or product references, and the only mention of Goodyear was as the program's sponsor.

Litchfield had become greatly concerned about moral laxity and indifference to human rights and dignity in the postwar era. In a letter to a business acquaintance, he expressed his feelings about sponsorship of this radio program.

> One person, two thousand years ago, confined to a radius of eight miles during his lifetime, traveling on foot or the back of a domestic animal, reaching only those within sound of his voice, left such an impression on the hearts of mankind throughout the world and over the centuries that we thought it would be a worthwhile contribution to society if this same lesson could be brought through the power of modern radio to a worldwide audience who, in such times as these, are so much in need of it.

A public service program to support agricultural conservation also was undertaken in 1947 at the instigation of Sales VP Wilson, who owned a small, unprofitable farm in Michigan and understood the need for scientific soil conservation. The project bolstered work in progress through federal, state, and local agencies and aimed to stimulate the saving of topsoil through planned scientific programs.

The Goodyear Soil Conservation Awards program was tested in eight Great Lakes and Corn Belt states for three years, then expanded into a national program that has been conducted annually ever since.

States are divided into districts of various geographic sizes, and winners of the top awards in this program are from two categories: an elected representative of the soil conservation district credited with the best accomplishment in his or her state, and the top farmer of that district. They receive plaques and are winter guests of the company for several days at Goodyear Farms and the Wigwam resort in Arizona.

Goodyear's tire and industrial product operations rolled smoothly along in the postwar years, but Goodyear Aircraft struggled. By May 1946, GAC's Akron payroll had dwindled to less than 1,000 hourly workers and about 1,200 salaried. The government-owned plant of the Arizona subsidiary had been completely phased out.

GAC continued, but at a slow pace compared with the hectic war days. It turned out aircraft canopies, airplane wheels and brakes, and a myriad of diverse items that included cabinets for refrigerators, electric freezers, jukeboxes, and 55,000 steel caskets for returning the war dead from temporary graves overseas.

Nevertheless, these contracts helped employment climb to more than 3,000 in early 1947, and in the next year contracts were signed to manufacture helicopter rotor blades, ice storage bins, washing machine subassemblies, and steel cabinets for sinks. In September 1948, the Navy's Bureau of Aeronautics awarded a contract to GAC for the engineering design of a new

*Workers at GAC's Akron factory building blimp gondolas. Goodyear repurchased several airships from the Navy after World War II's end for use in advertising and promotional campaigns.*

kind of long-range patrol blimp designated the N-ship. It was the largest patrol blimp ever contemplated at that time — 1,110,000 cubic feet — and was the predecessor of a long line of postwar airships manufactured for the Navy in the next decade.

Through an interest in missiles and their guidance, GAC ventured into the field of electronics, and an electronic analog computer its engineers developed to speed up their work gained a lot of attention outside the company. This successful development eventually led to GAC's manufacturing of computers, a product that would open the doors to a sizable portion of its future business.

As motor vehicle transportation burgeoned around the world, Goodyear concentrated foreign expansion on areas with high-growth potential. The company established tire factories in Colombia and Venezuela in 1945 and one near Havana, Cuba, the following year.

The first Goodyear tire produced in Africa came off the line in January 1947 at a tire and mechanical goods plant in Uitenhague, about 50 miles from South Africa's automotive manufacturing center of Port Elizabeth.

## A Factory Lost, Regained

On the other side of the world, in Indonesia, Goodyear-made tires once again rolled out of the factory at Bogor. Five months after Java's capture in 1941, the Japanese Army had begun operating the Bogor factory under the Nippon Tire Co.

Nippon Tire had been founded in 1931 as Bridgestone Tire Co., Ltd., but changed to the more nationalistic name of Nippon in 1942. In 1951, it returned to the Bridgestone name, and the Japanese company, which in 30 years would be the world's sixth largest producer of rubber products, signed a contract with Goodyear for technical assistance in tire development and manufacturing.

During the war, Nippon made nearly a half-million automobile, truck, and bicycle tires at Bogor. Although some Japanese characters were cut into the molds, the tires still carried the Goodyear name in English. The Japanese abandoned this factory in 1945 during the last months of the war, and area natives found it a choice place to ransack. Goodyear moved back in late 1946 to resume full production, and by the end of the following January nearly all former employees had returned.

Although Akron might have been considered parochial in the mid-1940s, the men who led its tire companies during the war had developed true global perspectives. They considered the world their market and moved into it boldly.

A new generation of Americans also raised its sights and extended its interests to far-off horizons. U. S. business began to move capital, technology, goods, and management skills across international borders with the same speed and purpose used to cross state lines in prewar times. Goodyear, far ahead in the tire and rubber industry, had by now put down roots in every inhabited continent, and international operations were gaining momentum.

## Golden Jubilee

Goodyear managers from around the world came to Akron for the company's Golden Jubilee Anniversary, celebrated on October 6, 7, and 8 in 1948. The festivities included pageants on the corporate history, presentation of the first 50-year service pins — to pioneer employees Al Cunnington and George Swartz — the premiere of a Goodyear motion picture, *Letter From America,* that focused on opportunities inherent in the free enterprise system, and production of the company's 450 millionth pneumatic tire.

In his keynote address, President Thomas emphasized the role of people at all levels in the company's progress and asked Goodyearites to respect past success as they planned and worked for greater success in the future.

"We look to the past for the lessons learned from our rich experience," he said, "but our eyes are on the future. We will have expanding markets for all the products we now manufacture . . . The answer to our relative success rests in the human side of Goodyear, in its people, its teamwork, and its spirit. That is the great heritage that has been bequeathed to us."

Chairman Litchfield said in a full-page message to Akron published in the *Beacon Journal*: "Akron and Goodyear have come a long way together in the past 50 years, but as the legend at the entrance of our Research Building promises — 'The Best Is Yet to Come.' "

As Goodyear marked its first half century, Litchfield, as always, spoke with prescience and authority.

# THE WINGFOOT CLAN

GOODYEAR

HOMECOMING EDITION

PROTECT OUR GOOD NAME

Vol. 37 AKRON, OHIO, WEDNESDAY, OCTOBER 6, 1948 No. 41

# 50TH ANNIVERSARY OPENS WITH LITCHFIELD GREETINGS

## Mayor Slusser Issues Jubilee Proclamation

In a proclamation issued today, Mayor-Manager Charles E. Slusser of the City of Akron welcomed the visitors from all parts of the world who are attending the Homecoming. The proclamation text follows in full.

PROCLAMATION

WHEREAS the Goodyear Tire & Rubber Company is celebrating its 50th Birthday Anniversary on October 6, 7, and 8; and

WHEREAS some 1,700 company executives will attend this Golden Anniversary from all parts of the world, which will serve as a forerunner for similar events to be hold in 58 cities in the United States where Goodyear has production or sales operations, as well as all foreign lands where Goodyear has representation; and

WHEREAS 50 years ago a small group of men were instrumental in building two bicycle tires, which officially placed the Goodyear Tire & Rubber Company in business. This was the forerunner for the world's largest rubber industry, and provided a nucleus for Goodyear's world-wide operations today; and

WHEREAS the personnel of the Goodyear Tire & Rubber Company now numbers 72,000 men and women around the world, with an ever-expanding production, covering literally hundreds of rubber products and other goods, making the company one of Akron's largest employers.

NOW, THEREFORE, I, Charles E. Slusser, Mayor-Manager of the City of Akron, Ohio, call upon the citizens of Akron to help celebrate the Golden Jubilee of the Goodyear Tire & Rubber Company. During this celebration Goodyear will produce its 450 millionth tire. Few industries can match the accomplishments scored in the rubber manufacturing field, as Goodyear passes the 50-year mark.

IN WITNESS WHEREOF, I have hereunto set my hand and caused the seal of the City of Akron, Ohio, to be affixed on this 30th day of September, 1948.

Chas. E. Slusser,
Mayor-Manager

### It Means Same Thing

In South America they say, "NO FUMAR EN EL TEATRO"

Sweden would pronounce it: "ROKNING FORBJUIEN I TEATERN"

Japanese say: "NIET ROOKEN IN HET THEATER"

Denmark would request — "RYGNING FORBUDT I TEATERET"

Our insurance company demands "NO SMOKING IN THE THEATER"

RETIRED AIR FORCE CHIEF OF STAFF IS GUEST

GEN. CARL SPAATZ

## Gen. Carl Spaatz Is Dinner Speaker

Gen. Carl Spaatz, retired Chief of Staff for the United States Air Force, will be the speaker tonight at the Homecoming banquet being held in Goodyear gymnasium commemorating the 50th anniversary of the Goodyear Tire and Rubber Company.

The wartime commander of the U. S. Air Forces in the European Theater of Operations and later commander of the U. S. Strategic Air Forces in the Pacific will tell the group of 1700 Goodyear executives and special guests of the duty of productive industry in times of national emergency. He will base his remarks on a wide-background of experience in first line combat through the two world wars.

A graduate of West Point with the class of 1914, Gen. Spaatz is holder of the Distinguished Service Cross, highest Army Award, for heroism in action in the St. Mihiel and Meuse-Argonne offensives as a fighter pilot in the Second Pursuit Group. He directed the famous 8th Air Force in its systematic destructive bombing of the German war machine and the final stages of the B-29 raids against Japan, including the dropping of the first two atomic bombs.

E. J. Thomas will serve as master of ceremonies for the banquet program. Miss Ethel Richardson,

(Turn to Spaatz, Page 2)

THE GREATEST NAME IN RUBBER
1898
50
GOODYEAR ANNIVERSARY
1948

## 1700 To Take Part In 3-Day Gala Program

The gang's all here this morning!

The grandest bunch of guys, about 1,700 of 'em from almost every country on the face of the earth, from the four corners of these United States, and from the home plants and offices, are gathering today for a three-day celebration of Goodyear's Golden Anniversary.

Almost every train, plane and bus arriving in Akron for the past week has brought early arrivals to the city, with the bulk of them coming in last night and this morning.

Not since before the war has there been any resemblance of such a reunion of Goodyearites. The last time the company called its world-wide representatives together was in 1939, in commemoration of the 100th anniversary of the Charles Goodyear discovery of the vulcanization of rubber.

The Mayflower, Portage and Akron hotels are actually bulging at the seams this morning. Their attractively decorated lobbies are jammed with new friends and old friends. It almost looks like a Cleveland Indians crowd at the World Series.

SEE COMPLETE PROGRAM PAGE 2

The city of Akron has extended its official welcome by a proclamation from Mayor Charles E. Slusser.

And that's not all!

**Instead of rolling out the conventional plush carpet, reserved for dignitaries, the city fathers went much farther for Goodyear's birthday party. The city has actually constructed a street of rubber, Goodyear rubber of course. Exchange street, from Broadway on the east to Five Points on the west, has been surfaced with a combination of synthetic rubber and asphalt.**

Goodyear rubber experts and highway construction authorities feel confident this stretch of highway will do more than a creditable job; possibly opening up undreamed potentials for this type of material in the years to come.

The Homecoming program officially gets underway at 10 o'clock this morning at Goodyear Theater. Harry Carroll, general traffic manager, and his capable lieutenant, Clyde Martin, have a fleet of 18 buses, appropriately marked, for transportation between downtown and Goodyear Hall.

**Buses will leave the three downtown hotels every few minutes, starting at 9 o'clock, and continuing until the program gets underway.**

**Goodyearites registered at the downtown hotels should report to their headquarters' room at their hotel this morning before leaving for the hall. Kits containing credentials, banquet and entertainment tickets, badge, program and other material will be distributed there. Akron personnel taking part received their credentials Monday and today.**

OPENS SESSIONS

P. W. LITCHFIELD

Caps and canes will be distributed at the close of tomorrow morning's sessions, before a group picture is taken at Seiberling field.

Highlights of today's program include greetings from P. W. Litchfield, chairman of the board and chief executive officer of the company, who, in his 48 years' service with the company, has seen it expand from a once-abandoned strawboard factory—just back of where the general offices now stand—to the world's largest rubber manufacturing company.

Tributes during the day will be paid to F. A. Seiberling, founder of the Goodyear company 50 years ago, who today is celebrating his 89th birthday. Despite his advanced years, there is a great possibility that Mr. Seiberling will

(Turn to Anniversary, Page 2)

*The Akron "Wingfoot Clan" of October 6, 1948, reflected the pride and optimism demonstrated in the three-day celebration of Goodyear's golden anniversary: pride in the company's wartime accomplishments; optimism in looking to a future of peace and progress.*

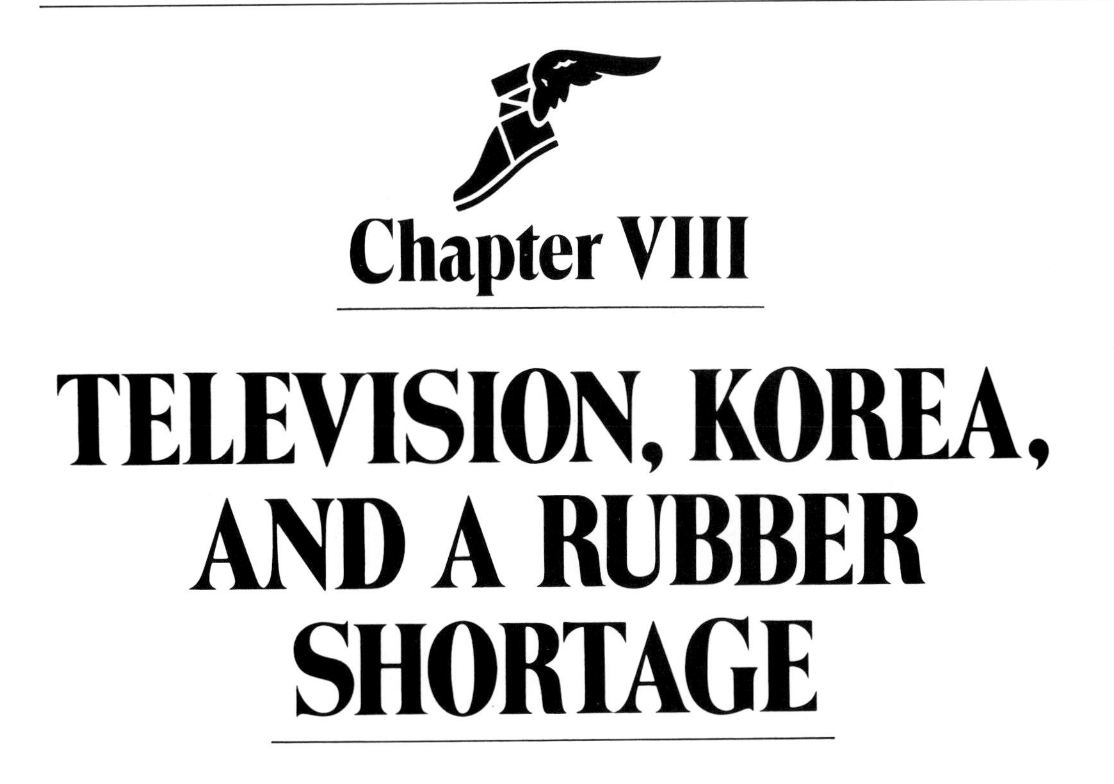

# Chapter VIII

# TELEVISION, KOREA, AND A RUBBER SHORTAGE

The staff of the *Wingfoot Clan* was especially busy during early October 1948. For the first time in its 37-year history, the *Clan* was published on successive days, October 6 and 7, and both extra-large issues focused on one story: Goodyear's 50th anniversary.

A banner headline on the sixth proclaimed: "1,700 To Take Part in 3-Day Gala Program."

"The gang's all here this morning!" began the story. "The grandest bunch of guys, about 1,700 of 'em from almost every country on the face of the earth, from the four corners of these United States, and from the home plants and offices, are gathering today for a three-day celebration of Goodyear's Golden Anniversary."

One photo showed a blimp towing a banner that read, "Welcome Home Men of Goodyear," and an extensive interview with founder F. A. Seiberling, "89 today," covered a half-page of the tabloid-size paper. The next day's *Clan* carried a two-page panoramic photo of the celebrants, all wearing Goodyear caps, seated on 14 rows of bleachers at Seiberling Field.

Speakers from outside the company reflected Goodyear's pride in its wartime accomplishments. Among them were Gen. Carl Spaatz, retired chief of staff for the U.S. Air Force, and Lt. Gen. Benjamin W. Chidlaw, deputy commanding general of the Air Material Command. Nevertheless, continued peace, the primary concern of a world debilitated by the most widespread and calamitous war ever, was in the hopes and outlooks of Goodyearites gathering in Akron to celebrate the past and consider the future.

President Thomas recounted the company's origins and early growth.

> From a simple beginning has come the great Goodyear company of today, a typical example of what is possible under the American free competitive enterprise system.
>
> In this 50 years we have taken in from our customers approximately $9 billion of revenues . . . $2.6 billion went for wages and salaries . . . over $4 billion went for materials . . . $720 million went to the government for taxes . . . $270 million went for plant replacements . . . $755 million went for operating expenses . . . and finally, $193 million went to those thousands who, over the years, provided the capital needed for growth of the institution, and $244 million was retained in the business to provide for growth.

> We have constantly passed on to customers the improvements and savings we have been able to make . . . The best illustration of this is that in 1910 a tire, size 30 x 3½ inches, cost $31.20 and gave 2,000 to 3,000 miles of service. Compare this today with an average passenger tire, size 6.00-16 inches, costing $15.95 and giving an average of 30,000 miles. No wonder the public has favored us with the purchase of 450 million tires in this 50 years.

He also expressed the hopes of a war-weary country.

> . . . The world situation looks bad in many respects, but peace is what all the people of the world want more than anything else, and we must do everything in our power to bring it about. America is the leading power in the world today, and we want to use that power only for good. The American people will sacrifice to help their neighbors, will help anyone who will contribute to a peaceful, productive world.

In closing the event, P. W. Litchfield asked celebrants to "think of Goodyear as one great family." He reminded them that "we are now entering the Air Age" and, echoing convictions expressed in the '20s and '30s, said, "Goodyear's future lies in becoming as closely identified with the progress of airway transportation as it has with highway transportation."

He further voiced the philosophy that was part of his legacy to the company he had led for three decades.

> In former years most businesses were started by an individual, grew and prospered as he lived and worked, and passed out of existence when he passed away. But in modern times the great plants for production and distribution require the investment of savings of thousands of people and gathering of other thousands in a never-ending procession to ensure security of the savings and the permanency of opportunity to reap progressively the rewards of one's skill and workmanship.
>
> The modern corporation therefore is not the magnification of a man, but the magnification of a family that, if successful, goes on from generation to generation. So we see Goodyear today as a great family of more than 70,000 men and women of all ages, scattered in large and small groups over all parts of the earth, striving for a common purpose: to make life better for those who buy the products of its labors and, as a natural by-product thereof, to reap the rewards of service well rendered.

Once again he discussed the Goodyear team approach to success, pointing out that it was

> injected into the Goodyear family in its early days. The training of the head and hand can go far, but only when the appeal to the heart is present can real greatness be achieved. Only by holding fast to the Goodyear spirit can we hope to keep Goodyear the greatest name in rubber.

## Commitment to Progress

Many of the conferees would return to their foreign posts and become direct participants in Goodyear's most dramatic international expansion, which had even then begun. They would take part in the economic resurgence of Europe and Japan after nearly a decade of the havoc of war and in the birth pangs of industry in many underdeveloped countries.

Undoubtedly the positive attitudes of these international Goodyearites reflected the brash confidence of the United States, which with its industrial might had turned the tide of global war. Conscious too of their nation's new responsibility toward a crippled world, they may have felt a stronger commitment than their predecessors to the economic and social progress of the countries in which they served.

Goodyear's 50th year produced a peacetime sales record: $705 million. About $204 million was paid out in wages and salaries to 72,000 employees, and nearly $12 million in dividends went to 45,000 shareholders.

The annual report, cautious about the future, reminded stockholders that "pipelines have been filled and inventories of manufacturers, dealers, and consumers are now ample." Management foresaw a "return to the buyer's market" and much keener competition than in the '40s, but the chairman's letter added, "It is under such conditions that our company made its greatest gains and built up public preference for its products."

Five months after the golden anniversary, death claimed Cliff Slusser, who exemplified early Goodyear managers. Dedicated, driving, extremely energetic, he had caught Litchfield's attention while leading a factory clean-up after the Little Cuyahoga's disastrous flood in 1913.

Like most early executive team members, Slusser had preferred to manage personally in the midst of action. He maintained close relations with many factory employees to the end of his career and was addressed as Cliff by workers in all the plants, even as a member of the board.

One of his prime goals had been to keep the Goodyear family spirit strong throughout the ranks despite the rapid expansion that had tended to separate management from factory employees.

## Pioneer Advertising

The company's 1948 concerns were fully justified in the first half of 1949. Six-month sales dropped $35 million, or 10 percent, and profits fell 43 percent to only $8.1 million, $6 million below the same period in 1948.

A news release reporting the first-half results of 1949 blamed the poor showing on "increased costs of doing business all along the line and lower prices resulting from the return of intense competition in the tire industry."

As always, Goodyear met this competition head-on and was well armed to do so. Its working-capital position was strong — $256 million versus $249 million in 1948. The ratio of current assets to current liabilities was 6 to 1, and management had full confidence in its product line.

Aggressive marketing for the 1950s began on November 6, 1949, with the sponsorship of a weekly television program, "The Paul Whiteman Revue," on ABC television. Goodyear had been a pioneer sponsor on network TV three years earlier, televising the West Point football games on NBC stations connected by coaxial cable; at that time the entire United States had only 9 TV stations and 34,000 receiving TV sets.

By 1949 and sponsorship of the Whiteman show, stations numbered 87 and American home sets 3 million. Goodyear sponsored this program three seasons and was a leading TV advertiser throughout the '50s.

In 1951, the company cosponsored television's top one-hour show, "TV Playhouse," with Philco on an alternate-week basis. It was telecast by 57 NBC stations. The company maintained its religious noncommercial radio

*1948 ad in "Collier's," "Saturday Evening Post," and "Life."*

series, "The Greatest Story Ever Told," through 1956 and occasionally telecast it over either the NBC or ABC network.

Goodyear's three L-type blimps continued to operate for advertising and promotion purposes and were becoming known as company symbols.

Although volume blimp building was now part of Goodyear's wartime history, the Navy continued to operate airships for coastal patrol and in early 1949 awarded a contract to the company for manufacture of the world's largest blimps. Construction began in August on the new airships, designated the ZPN-1, or N model, which incorporated the Navy's latest technology for locating enemy submarines.

The N ship was 324 feet long, 94.5 high, and 73.5 wide. Its capacity was 875,000 cubic feet of helium, compared with 725,000 for what had been the largest blimp, the M model, and 456,000 for K-type ships, of which more than 130 had been used in World War II antisubmarine service. ZPN-1 carried a crew of 14 and could refuel in flight from surface craft.

The prototype of the new ZPN-1 blimp class made a successful initial test flight on June 20, 1951, and was delivered to the Navy at Lakehurst on June 17, 1952.

## Scrambling for Business

For Goodyear Aircraft Corporation, the transition from wartime to peacetime was completed quickly, and management began an eager search for customers. Early in 1949, Goodyear purchased the idle aircraft plant in Litchfield Park, Arizona, at public auction for $475,000 from the War Assets Administration. One of its first jobs was to make the large envelope required for ZPN-1; from then on all the company's airship envelopes were made at Litchfield Park.

Montgomery Ward & Co. contracted GAC to make steel cabinets for use in kitchens, but military items still made up most of its production. Air Force contracts called for fuel tanks and nose assemblies for the Boeing 50D bomber and gasoline tanks containing fuel cells for the big B-36 bomber. In cooperation with its parent company, GAC provided belts and brakes for Caterpillar-type landing gears tested on the B-36s. The Army Quartermaster Corps awarded a $1 million contract to GAC for steel lockers.

GAC's need to scramble for customers lessened in June 1950 when the United States became embroiled in the Korean War. The reputation Goodyear Aircraft had established for the high-quality production of airframes and plastic laminate aircraft components during World War II proved an asset in gaining important subcontracts during this conflict.

Because the Korean engagement was a "limited war," it required only a small diversion of the nation's industrial might to military production. Many wartime manufacturers had leftover facilities with which to make aircraft parts, so competition for subcontracts was intense.

Thanks to its reputation, its wide-ranging capability in military production, and an opportunistic sales staff, GAC got a fair share of business.

- Wings, empennage assemblies, and canopies for North American's T-28 Air Force trainer
- Stabilizers, elevators, and canopies for Republic's F-84 *Thunderjet* fighter-bomber
- Prefabricated radar towers three stories high for General Electric's remote-area electronic devices
- Fuel tanks, canopies, radomes, and ducting for anti-icing systems for the Boeing B-47B bomber
- Canopies for Northrop's F-89D *Scorpion* jet interceptor
- Canopy sections and other parts for Beech Aircraft's twin-engine Air Force trainer-transport plane, the T-36A
- Nose antenna housings, fuel-tank liners, and canopies for Grumman's F9F *Panther* jet fighter
- Flooring, hatches, and other parts for the Navy's Grumman S2F-1, an antisubmarine search-and-attack aircraft
- Radomes for the Douglas C-124 *Globemaster*, the Air Force's giant cargo plane
- Fuel tanks for the C-97 *Stratofreighter*, used by the Air Force for the in-flight fueling of jet aircraft

During the Korean War, GAC also built the first successful autopilots for helicopters, designed by the Air Force, and for the Navy produced the working prototype of the escape capsule, an enclosure to protect pilots ejected from disabled aircraft at high altitudes or supersonic speeds.

## Into the Space Age

GAC had been working in a new field of military science — rocket and missile components — and in November 1951 broke ground for a building at the aerospace complex in Akron to house this classified production.

An augury of Goodyear's involvement in space technology appeared in 1952 with the announcement of the L3 GEDA, Goodyear Aircraft's improved version of its electronic differential analyzer, popularly known as the electronic brain, or computer.

In describing this new development five years before Russia launched *Sputnik,* the *Wingfoot Clan* reported:

> Flight engineers, faced with the weighty problems confronting builders of the world's first spaceship and construction of a satellite body outside the earth's atmosphere, will be greatly aided in their pioneer efforts by the L3 GEDA . . . With the L3 GEDA and related facilities in applied dynamics laboratories, many problems of space flight can be solved. The actual rocket ship can be simulated in the GEDA, the various flight factors being translated into mathematical equations and solutions without actual trial and error test flights.

As the sea-lanes reopened and military demand for rubber dropped sharply after World War II, a balance seemed to occur in the production of natural and synthetic rubber. Throughout 1949, general purpose synthetic rubber hovered around 18¢ a pound. Natural rubber dropped to about 16.5¢ in midyear, but rose to 19¢ by year's end.

Assuming the nation's rubber supply would be steady and secure, the government shut down some of its synthetic plants, including the Goodyear operation in Akron. Government agencies failed to increase stockpiles of natural rubber, however, when it was selling at less than synthetic rubber; they further neglected to build an adequate reserve of synthetic.

Goodyear continued to operate the government's Houston synthetic plant and in 1948 changed some of its production to the continuous cold-rubber process. In this process, rubber is manufactured at 41°F — in contrast with 122°F at which conventional synthetic GRS is produced — and is a tougher, stronger material.

The company regained its Dolok Merangir rubber plantation in Indonesia in 1949. About one third of its original trees had been destroyed during the Japanese occupation, but more than 10,000 of the 20,000 plantation acres remained in good condition.

## A Crisis in Rubber

Rubber's price structure crumbled in 1950 when demand far exceeded supply; this situation developed because of both the Korean War and record production of cars and trucks in the United States and other auto-manufacturing countries. Political unrest in Southeast Asia's rubber-producing countries intensified the need for an increased rubber stockpile in the United States, and by the end of the year, the price of natural rubber had skyrocketed to 80¢ a pound; it exceeded 81¢ in early 1951.

Long before the rubber situation had become critical, Litchfield warned that the nation's emergency stockpile had dwindled to a dangerous low and that synthetic rubber production had been cut too much. Just before the outbreak of the Korean conflict, he issued one of a series of publications entitled *Notes on America's Rubber Industry,* in which he wrote:

> There are two things radically wrong in the rubber situation today: 1. Our supply of raw rubber is too low from the standpoint of national security. 2. The price of natural rubber is too high for economic stability.
>
> Both situations trace primarily to our national rubber policy, once so clear, decisive and effective, but now seemingly weakened by a creeping uncertainty of purpose. The remedies are immediately available if we will only apply them. They are: 1. Stepping up our output of synthetic rubber to 50,000 long tons per month from the present output of 40,000 tons. 2. Building up an inventory or stockpile of synthetic rubber to a minimum of 200,000 long tons.

On August 25, 1950, two months after the start of U.S. involvement in Korea, Washington moved to ease the rubber crisis. Federal controls were reestablished, and the tire and rubber industry's consumption of rubber — both natural and synthetic — was cut back 20 percent.

The government-owned and Goodyear-run plant in Akron, built in 1942 to assist in the war effort and closed in 1947, was reactivated with several other idle synthetic rubber plants. By 1951, Goodyear factories processed more than one seventh of the world's total consumption of new rubber annually, and Goodyear operated the government's synthetic plant in Houston, the world's largest producer of man-made rubber with annual capacity of 200,000 tons.

A return to large-scale synthetic production and a rebuilding of the national rubber stockpile soon led to the restabilization of natural rubber prices. Government controls were rescinded on May 1, 1952.

## Synthetic Catches On

Consumption of rubber in the United States — both synthetic and natural — reached 1.26 million long tons that year, a record high. This was about double the consumption of 1940, the last full year before World War II, and was equivalent to 18 pounds per capita. Consumption for the rest of the world was slightly more than one pound per capita.

The price of high-quality natural rubber in 1952 dropped from 49¢ a pound in January to a low of 27¢ in October, averaging just under 34¢ for the year. The government's selling price for synthetic rubber was pegged at 26¢ a pound through February, then reduced to 23¢ where it stayed for several years. This time, though, new rubber products, such as foam rubber, film, plastics, and industrial items, accounted for a third of all rubber consumed.

U.S. rubber consumption was divided almost equally between natural and synthetic in the mid-1950s. In the rest of the world, consumption of natural was 90 percent, synthetic only 10 percent. World consumption of natural was about two thirds of the total.

In 1954, Goodyear acquired its first new plantation in almost 20 years by purchasing 11,000 acres near Belem, Brazil. It was called the Marathon Estate and was totally undeveloped. The Brazilian government, hoping to increase its domestic supply of natural rubber, assisted in the plantation's development.

*Opposite, tractor-trailer fitted with tires made from Goodyear's new synthetic rubber, polyisoprene, for the Natsyn test run of 1957. The cross-country journey demonstrated company claims that the new product produced tires equal in quality and durability to natural rubber tires. Research in the development of synthetic rubber was stimulated by a critical rubber shortage in the early 1950s and political instability in the rubber-producing countries of Southeast Asia.*

A year later, after Congress had authorized the sale of government-owned synthetic rubber plants, Goodyear bid successfully for the two it operated in Akron and Houston. The sales were made possible by the Rubber Producing Facilities Disposal Act of 1953 and congressional conviction that rubber shortages were unlikely in the future. The Akron plant was bought for a little more than $2 million, Houston for $11.9 million.

The synthetic rubber plants purchased from the government by Goodyear and other rubber manufacturers increased total industry capacity by about 500,000 tons annually in the first three years of private ownership. By the end of 1957, the Houston and Akron plants together turned out 247,000 tons a year, making Goodyear the world's largest producer of man-made rubber.

Goodyear-owned sources of natural rubber in the mid-1950s included 55,000 acres of rubber plantations in Sumatra, 2,500 acres in both the Philippines and Costa Rica, 11,000 acres under development in Brazil, and a new experimental plantation of 3,400 acres in Guatemala, acquired in 1957.

The company's rubber researchers and development engineers kept pace with production gains. In 1955, they created a synthetic rubber — polyisoprene — that nearly duplicated the molecular properties of tree-grown rubber. For commercial purposes it was named Natsyn, combining the first syllables of natural and synthetic. A pilot plant was set up for small-scale production of this new product in 1956, and operations began in 1957.

To prove the quality of Natsyn, Goodyear revived the highway demonstration technique begun in 1917 when the Wingfoot Express demonstrated the capabilities of pneumatic tires on trucks in its Akron-Boston run. Natsyn got its first public trial on a tractor-trailer rig of the Goodyear test fleet. The rig was equipped with natural rubber tires on one side and Natsyns on the other.

The tires were switched regularly from one side to the other on a run at legal speeds from San Angelo, Texas, to California, then to the Atlantic Coast and back to San Angelo. Analyses at journey's end showed the Natsyns had equaled the natural rubber tires in all aspects of performance, including durability.

## The Proving Ground

The Natsyn test run had started at San Angelo because Goodyear's major test fleet operations had been headquartered there since 1944. Before then nearly all road testing was conducted in the Akron area, with occasional highly publicized cross-country excursions, such as the Wingfoot Express and the Transcontinental Motor Express. In 1923, winter testing was conducted in Tampa, Florida, and continued there until 1931 when it was transferred to Litchfield Park.

By 1937, all test operations, year-round, were concentrated at the Arizona facility. Besides its good conditions for testing truck and passenger tires, the Goodyear farm properties were ideal for testing tractor and farm tires on the many vehicles in normal use there.

When the Phoenix-Tempe-Scottsdale area began to show rapid population gains and greatly increased highway traffic, a search for a less-congested area with a constant climate was begun, ending at San Angelo.

Eventually, though, the highways there also became too crowded for the constant testing that kept Goodyear's cars and trucks on the road 24 hours a day; three drivers worked 8-hour shifts and stopped only for periodic tire checks.

The answer was its own proving grounds on its own property; so in 1957 the company began construction of a 7,200-acre tire test site with its own highways near San Angelo. This multi-million-dollar center included a banked, high-speed circular roadway 5 miles around; an 8-mile paved course designed as a figure eight; a 2½-mile straightaway for truck-tractors; a 3-mile meandering gravel road; and headquarters buildings and garages. The 5-mile circular track was completed in 1958 and named "The Turnpike That Never Ends." It was engineered for speeds of up to 140 mph without effects from centrifugal force.

*E. J. Thomas, president, and P. W. Litchfield, chairman, with Goodyear's 650,000,000th tire, 1955. Litchfield handed over the reins of leadership in 1956, after five decades as a director and three as CEO of the company he had led to world preeminence.*

# Chapter IX

# THE FIRST BILLION-DOLLAR RUBBER COMPANY

In contrast with the turbulent 1930s and 1940s, which had included both the globe-shaking Great Depression and World War II, the '50s were placid for Americans despite a three-year involvement in the Korean War. After that conflict, the nation settled into the calm years under President Dwight D. Eisenhower, content with healthy economic growth at home and expansion of its economic interests abroad.

Goodyear's sixth decade, 1949 through 1958, was a period of dynamic, almost hectic, progress both in the U. S. and in international markets. Growth was tremendous. In its first half century, total sales had been about $9 billion; in this decade alone they exceeded $11.5 billion.

On February 18, 1952, the national press reported that in 1951, Goodyear became the first rubber company to exceed a billion dollars in annual sales, with $1,101,141,000, a 30 percent increase over 1950.

Consolidated net income that record year amounted to $36 million, equivalent to $8.18 per share on more than four million outstanding common shares following a two-for-one split. Sales for 1958 topped $1,367.5 million; net income was $65.7 million.

Income from foreign subsidiaries in 1951 was $17.2 million. By 1958, it had risen to $18.5 million, only a hint of the international growth ahead.

In more than half a century as a tire and rubber company, Goodyear had made several major ventures into seemingly nonrelated fields that in fact were closely tied to the company's main business. Its cotton mills and cotton plantations provided fibers for tire making, and before building dirigibles and blimps, it had supplied rubberized fabric and tires to airplane builders, including the Wright brothers. For the most part, all company products were directly related to the burgeoning transportation industry.

Goodyear stepped into a new world in 1952, the world of nuclear science, a venture begun by a telephone call from the Atomic Energy Commission (AEC) to the company's president.

"Would Goodyear be interested in discussing the possibility of taking on a major government project?" Although no explanation was offered, E. J. Thomas replied, "Yes. We would like to discuss it."

*Atomic power plant in Portsmouth, Ohio. Goodyear entered the burgeoning field of atomic energy in 1952 when chosen by the Atomic Energy Commission to operate the immense new $1.2-billion southern Ohio facility.*

## Goodyear Goes Nuclear

In an Akron conference, he and other company executives learned on a confidential basis that the AEC sought a company to operate an immense plant under construction in southern Ohio that would produce concentrated uranium 235 isotopes by the gaseous diffusion process. The $1.2 billion facility would be completed in four years.

Enthusiastic about a possible entry into the field of atomic energy — several companies were vying for this opportunity — and impressed by the project's magnitude, Thomas personally represented Goodyear in all prime negotiations with the AEC. One of the first exploratory steps was a group visit by several top executives to the Oak Ridge, Tennessee, atomic plant operated by the Union Carbide Nuclear Corp.

All preliminary negotiations were conducted under tight security, and after many weeks, Thomas was told competition had been reduced to five companies. A few weeks later, he received word that Goodyear had been chosen for the assignment; public announcement would be made on September 18, 1952. Thomas asked for, and received, permission to inform employees a half hour before the official news release.

On that eventful day, 2,000 people gathered in Goodyear Theater and were told the news at the same time Goodyearites in plants and sales offices across the country were being informed.

"It will be part of Goodyear's job to become thoroughly acquainted with this new plant and its equipment as it is erected," Thomas said, "during the construction period to train personnel who can undertake its operation as soon as any given portion of the plant is ready, and to test the plant and its equipment to see that it will function as planned."

He gave three reasons why the board had unanimously supported the undertaking.

"One, we feel that it is the duty of a company such as Goodyear to take on such an obligation in the interest of the defense needs of the nation.

"Two, we feel that an aggressive company should be allied with this important development.

"And three, we feel that it is a good thing for the organization to accept such a challenge."

A Goodyear subsidiary, Goodyear Atomic Corp., was established to conduct the project, and its officers were principals of the Goodyear company. Thomas reported, "The responsibility for the operation ties into the parent company through Russell DeYoung, vice-president in charge of production." He also announced that A.J. Gracia, whose work in synthetic rubber research had gained wide recognition, would be plant manager.

## $450 Million Savings

Well ahead of schedule, the plant's first unit went on stream in September 1954, just two years after Goodyear was selected. The plant's cost of $750 million was more than $450 million less than the original estimate.

Goodyearites soon learned the immensity of the new subsidiary. Its 4,000-acre site was enclosed by 11 miles of fence, and a 3.7-mile stretch of fence surrounded the actual production area. Floor space of the permanent buildings totaled 220 acres, and each of the three main buildings was about a half-mile long and 80 feet high. The processing equipment included 90,000 recording instruments, 600 miles of pipe, 1,000 miles of copper tubing, and 4,600 miles of wire and cable.

Original water requirements were 345 million gallons; about 30 million gallons had to be replaced daily. Electric current required burning 7.5 million tons of coal a year, and the annual power bill in the early years was about $72 million. The total power consumed by the plant represented two thirds of all required in the State of Ohio and three percent of the total generated in the United States.

Al Gracia explained the plant's function simply. "Ten tons of raw material going into the system for processing on any given day will yield something less than one quart of the desired product, U-235, at the other end of the system three months later."

## Expansion on All Fronts

Capital expenditures in the 1949-58 period for expansion and modernization of existing factories and establishment of new ones in the United States and abroad was about a half-billion dollars. In 1948, capital outlay at home was low, mainly because of the tire industry's rush to return to full peacetime production and a slump in the economy. The auto and tire industries on other continents, however, were booming; tire factories there were operating at capacity to meet demand. American tire makers took heed.

The U.S. economy picked up in late 1949, and prospects seemed bright for increased auto and truck production, both at home and abroad. The motoring population was exploding, and more and bigger highways were threading smoother ties between cities.

Goodyear perceived some good years ahead and embarked on major worldwide expansion programs. Most of the company's U.S. manufacturing installations benefited during this 10-year span, in some instances at costs exceeding original construction expenses.

The domestic buildup began with a 25 percent increase of capacity at Topeka for truck tires and tires for other large vehicles. The Akron Chemigum plant was expanded by 50 percent, and at Niagara Falls, New York, the vinyl resin facility increased production by 100 percent. One of the most significant expansions was at Gadsden where $11.5 million went into improvements and a new truck-tire factory, making this the largest tire manufacturing facility in the United States.

In 1955, Goodyear bought an idle plant in North Chicago, Illinois, to produce hose and other industrial products. A year later, the first Goodyear facility for making tread rubber was built at Chehalis, Washington, and a new $5 million distribution center and warehouse at Brook Park, near Cleveland, began operations.

Industrial demand for rubber products stimulated hectic activity in the company's nontire operations as well. In December 1949, the company installed the world's longest single flight of conveyor belting — 22,000 feet of steel cord belting — for the Weirton Steel Co., near Morgantown, West Virginia. It delivered 250 tons of coal per hour to barges on the Monongahela River through a tunnel from the mine area.

## A Big Lift

Little more than one year later, Goodyear supplied conveyor belting for the highest lift conveyor in the world. It moved 1,200 tons of coal an hour over nearly 2 miles at a 16-degree incline for a vertical lift of 862 feet at the Waltonville, Illinois, mine of the Chicago, Wilmington & Franklin Coal Co. Later in 1951, an even more dramatic belting system was installed to transport iron ore from 1,600 feet under the North Atlantic at the Wabana mine of Dominion Steel & Coal Co., Ltd., on Bell Island, Newfoundland. The belting was manufactured by Goodyear Canada.

While developing advanced products for the mining industry, Goodyear closed its own mining operation. In 1949 when its deposits became depleted, the Wheeling Township Coal Mining Co., near Adena, Ohio, was sold after 30 years of steady mining.

Much of the impetus for Goodyear's progress with industrial products in the 1950s was the result of new ideas in machine and equipment design. An influx of engineers from leading educational institutions joined forces with pragmatic old-timers, who had gained most of their knowledge on the factory floor, to keep the innovative process moving at a rapid pace.

The influence of the older employees was especially recognized in May 1954 when the Goodyear-developed "world's largest conveyor belt press" was unveiled in factory ceremonies as the "Judge Foster Press." Foster, who had joined the company in 1914 and was manager of the Industrial Products Division, had been instrumental in developing and installing this huge machine.

A few days later, the world's first passenger-belt conveyor, designed by Goodyear and Stephens Adamson Manufacturing Co., went into operation in Jersey City, New Jersey. This moving sidewalk was operated by the Hudson & Manhattan Railroad to carry passengers from the New Jersey end of tunnel tubes under the Hudson River up an incline ramp 227 feet long to the Erie Railroad Terminal.

*Top, patrons of the Cleveland-Hopkins Airport are spared up to 445 feet of walking between airplanes and the parking garage by two Speedwalk passenger conveyor systems built by Goodyear. Bottom, a youthful shopper tries out one of two moving sidewalks, which feature ribbed belts for maximum passenger safety, at the Mission Valley Center in San Diego.*

Goodyear introduced rubber material for railroad crossings in 1954. The first installation was at the Erie Railroad's intersection with U.S. Highway 42 at West Salem, Ohio, in November 1955. In the same year, Goodyear research engineers developed air springs for autos. Springs of this kind had been in use on buses, trucks, and trailers, but they had not been applied to automobiles because of the cost to owners of modifying vehicles designed for other kinds of suspension. In 1957, a few large automakers offered Goodyear air springs as optional equipment on new cars, and Plant II in Akron began producing them in volume.

During the flourishing economy of the '50s, Detroit factories hummed, car dealers grew wealthy, and the nation moved into its "Autopian Age."

## Resurging Foreign Markets

Bicycles, scooters, and motorcycles, however, were still major means of transportation overseas — even in developed nations. The streets of Amsterdam, Copenhagen, and Stockholm were crowded with bicycles, and American tourists in Rome seemed to be constantly dodging motor scooters on the crowded streets. With assistance from the Marshall Plan, which revitalized European markets, the economy across the Atlantic was fast recovering from the war, and the "economic miracles" of Italy and Germany were just around the corner. Europe was, in fact, taking the first steps into its own Auto Age.

In the developing countries, highways became key elements in growing infrastructures, and motor vehicles helped speed the growth of commerce, mining, and agriculture as they expanded the horizons of citizens. Foreign tire markets exerted a stronger pull than ever on U.S. tire manufacturers.

Goodyear had established a tire production beachhead in the United Kingdom and on the Continent with factories in Wolverhampton, England, built in 1927; the Stockholm plant, put into operation in 1938; and the Luxembourg facility, opened in 1951. Firestone had tire factories in England, Spain, Sweden, and Switzerland; and General produced tires in the Netherlands, Portugal, and Spain. The European tire markets, however, were dominated by home companies: Dunlop in the United Kingdom, Michelin in France, Continental and Metzler in Germany, Pirelli and Ceat in Italy, Semperit in Austria, and Englebert in Belgium. Principal competition came from among themselves and second-rank European tire companies.

Freight costs, high labor expenses, and national barriers, both tariff and nontariff, hampered tire exports to Europe. The same impediments inhibited gains in many other tire markets of the world. The idea of establishing tire production in much greater volume outside the United States thus became even more compelling.

## 'Global Surge'

Goodyear's drive into foreign tire markets gathered strong momentum in the '50s. In 1949, the company had made contract arrangements with Dunlop to make tires in New Zealand. Two years later, it organized a sales company in Japan, which was taking its first steps toward its own economic miracle that would astound the world by the late '60s.

Bridgestone Tire Co. would produce Goodyear tires there under an agreement by which Goodyear provided technical assistance. During World War II, Bridgestone — then Nippon Tire — had managed the Goodyear factory in Bogor, Indonesia, for the Japanese Army. John J. Hartz, manager of product development compounding, was sent to Japan as production manager to supervise Bridgestone's manufacture of Goodyear tires for Japanese markets. He was to become Goodyear's first vice-president for development in 1968.

The company completed arrangements in 1951 for the production of Goodyear tires in West Germany by Continental Gummiwerke at Hanover.

In June 1956, a new plant at Cali, Colombia, began operations, replacing a smaller one that had been producing tires since 1945. Goodyear's seventh factory in Latin America, at Valencia, Venezuela, was inaugurated in August, and in October of that year a plant was opened in Manila, in the Philippine Islands. A factory at Garscadden, near Glasgow, Scotland, began operations in September 1957. Construction started on two more tire plants in 1959, Amiens, France, and Medicine Hat, Alberta, Canada.

The primary motivation for Goodyear's global surge of the '50s was expressed by President Thomas at a Detroit press conference in December 1955.

"Great as are the prospects in the domestic markets for our goods," he said, "I believe the field of our foreign operations will probably expand even faster. When you stop to think that in this country there are about 20 pounds of rubber used per person per year, and the figure for the rest of the world is only a little over one pound, you can realize why I feel as I do on this point."

To emphasize Goodyear's commitment to foreign markets, he named the 17 countries in which the company had production facilities: Canada, England, Scotland, Sweden, Luxembourg, South Africa, Argentina, Brazil, Peru, Colombia, Venezuela, Cuba, Mexico, Australia, Java, the Philippines, and Sumatra.

## Changing Hands

Following World War II, the company's sales increased from $716.1 million in 1945 to $1,372.1 million only 10 years later. In August of that year, 1955, the man who had founded Goodyear and guided it to a sales volume of more than a half-million dollars in its first year of operations died at the age of 95. Frank A. Seiberling, the Little Napoleon of the tire industry who headed Goodyear during its first 23 years, was forced out in 1921, then founded the Seiberling Rubber Co., expired in Akron General Hospital only a few miles from the organization he had launched and led to world leadership.

Goodyear's top management changed hands for the first time in three decades on April 4, 1956. After 30 years as chief executive officer, Paul W. Litchfield stepped aside. Nearing 81, he continued as chairman of the board, but passed along his responsibilities and authorities as CEO to Edwin J. Thomas, who had been his secretary and assistant in the early 1920s.

These two men had shared many successes, disappointments, and hopes. They had grown with the company, fashioned its management of the '40s and '50s, and had similar perspectives of where Goodyear was going and how it would get there. But they were different personalities.

The quiet and dignified Litchfield had a father image and was well-known for the distinguished bearing he carried wherever he went.

Buoyant and articulate, Thomas was more outgoing and had a sense of timing, both in business affairs and public relations, that close associates described as perfect. One company executive once said: "When E.J. walked into a room with that bouncy step and ready smile, he took charge. He had a tremendously engaging personality, and he was totally liked and respected inside and outside the company."

Thomas had been Goodyear's first and only executive vice-president, serving in that capacity from 1937 until 1940. As the new CEO, he immediately announced the election of three new executive VPs: R. S. Wilson, sales; P. E. H. Leroy, finance and accounting; and Russell DeYoung, production, personnel, and research and development.

He also established a policy committee he headed as chairman, which included the executive VPs; Litchfield; Howard L. Hyde, vice-president and chief counsel; and Leland E. Spencer, a newly elected VP who had been assistant to the president. Victor Holt, Jr., and Francis J. Carter became corporate VPs, and two months later long-time VPs J. M. Linforth and Fred W. Climer retired.

## To Serve the Public

The Akron edition of the *Clan* recounted activities in the Goodyear Theater when 1,700 key personnel learned of the new management lineup and heard talks by Litchfield and Thomas. It described Litchfield as being "seated calmly on the great stage" when announcing the end of his career.

"Yesterday I completed my 50th year as a director of the company and 30th as chief executive officer," Litchfield said. "None of us can control age. I have been blessed with a long period of usefulness. In 1952 a change of health started me building for this day. I am sure Mr. Thomas will meet every problem. He has been thoroughly trained to carry on. I am confident of your new leadership."

At that point, the *Clan* reported, "President Thomas walked on stage and placed an arm about the shoulders of Mr. Litchfield, then strode confidently to the microphone on the opposite side of the stage to deliver his first message as chief executive officer . . ."

Litchfield was "one of the world's greatest industrial statesmen and one of the world's greatest humanitarians, whose first consideration was always the human factor," Thomas told the assembly.

> When I look out my window, as I sometimes do at quitting time, and see hundreds and thousands of people streaming from our plants and offices who are dependent on the success of our enterprise, and then think of the 100,000 employees plus their families in the entire company, and the 44,000 investors who have risked their money, and the 50,000 dealers and their employees and families, and the various governments that last year received more than $82 million from our company to help pay the costs of government: it is indeed a sobering thought, and it shows the dependence of so many people on how well you and I do our jobs.
>
> With all of this, we have the challenging responsibility of always trying to balance the best interests of customers, employees, shareholders, and suppliers. Our basic aim is to serve our boss, the consuming public, better.

Apprehensive, perhaps, that Litchfield might have difficulty in putting aside the old boss-assistant relationship, Thomas had earlier asked him if he was indeed handing over the reins.

"You will be the boss, Eddie," he said.

"And that's how it was," Thomas recalled many years later, "from the day I succeeded Litch."

## R & D in High Gear

Mindful of the increasing importance of research, development, and innovation in maintaining industry leadership, the Litchfield-Thomas team pushed hard for new and improved products in the 1950s. Goodyear research and development engineers responded, and their flow of inventions and improvements kept the sales forces hopping.

Among the most noteworthy of the new products was continuous-flow equipment for the 3-T process (time, temperature, and tension) of preparing tire fabrics. A Goodyear 3-T machine preconditioned tire cord — either nylon or rayon — before subjecting it to high tension and high temperature. The $1 million machines that processed the fabric were four stories high and 150 feet long at the base. Fabric traveled one third of a mile within a machine as it was processed.

A team of 20 researchers and a half-dozen development engineers — headed by G. D. Mallory, research division engineer, and Phil Drew, manager of fabric development — organized and conducted the chemistry, mechanics, production processes, and testing for the 3-T project. Mallory and Drew received Litchfield Special Awards of Merit in June 1955 for their work. One of the first new products made possible by the 3-T process was a full line of tubeless tires, introduced in mid-1954.

Tubeless tires were not new, however. A pneumatic tire patented in 1845 by Robert W. Thomson of Middlesex, England, was tubeless. Litchfield had been granted a U.S. patent for a pneumatic tubeless tire in 1903, and there were others in the early days of manufacturing.

Goodyear's tubeless, though, was a new concept, because it held air without a heavy, tubelike liner. Instead of a rubber liner or heavy puncture-resistant material to hold in the air, it depended on an airtight construction of the carcass.

When Thomas introduced the new tire at a Detroit press gathering in August 1954, he emphasized it had been made possible by the 3-T process of pretreating cord fabrics, a process exclusive to Goodyear. He announced that the 3-T principle was being used not only for passenger tires, but for truck, tractor, airplane, and bus tires as well. Two months later, the U.S. Air Force ordered Goodyear tubeless tires for the nose wheels of the Republic F-105 jet fighter. In January 1955, the Navy specified them for all wheels of its Grumman F9F-9 *Tiger* and North American FJ-4 *Fury* jet fighters.

For use with large tubeless tires, Tru-Seal rims were introduced for applications in which normal rims were impractical. By 1955, Goodyear was marketing tubeless tires weighing more than a ton for the largest earth-moving vehicles in operation.

Responding to fast-growing traffic safety consciousness in the United States during the '50s, the larger tire manufacturers worked hard to develop protection against blowouts. Goodyear's major candidates were the Nylon Lifeguard Blowout Shield and the Nylon Captive-Air Safety tire.

The Lifeguard Blowout Shield provided double air chamber protection in the top-of-the-line Double Eagle tire. The inner chamber of the tire-and-shield combination in the tire was inflated through the regular valve in the rim. The outer chamber took air through a self-sealing rubber valve in the tire sidewall, much like the valves in footballs and basketballs. The tires permitted a motorist to drive up to 100 miles after air in the outer chamber had escaped because of puncture or rupture.

As the car craze heightened in the '50s and long-haul trucking became a major industry, the tire industry became increasingly competitive; new and improved tires rolled out of development departments with furious frequency. Goodyear's most successful entries in the new-tire battle included the Suburbanite winter tire, the Super Soft Cushion ultralow pressure 14-inch tire, the Blue Streak Special, and the All-Weather Raceway and All-Weather Speedway — all passenger tires.

Among new Goodyear tires for trucks were the Hi-Miler Xtra Tred; the Unisteel, built with a single radial wire ply and a three-ply breaker belt; and the Hi-Miler Cross Rib. The company introduced the Traction Sure-Grip for farm tractor rear wheels and the Super Rib for front wheels. The first all-nylon bicycle tire, the Suburbanite 175, was introduced in 1957, the year the New Bedford factory produced its 30 millionth bicycle tire.

Although management pushed for a larger share of expanding tire markets worldwide, it could not deny the surging demands of industry and the public for nonautomotive rubber and plastic products. New nontire Goodyear products brought out in the '50s included Porolated vinyl film, similar to conventional vinyl film, but porous to permit ventilation, which won quick acceptance for rainwear, mattress and pillow covers, baby pants, and related products; a vinyl floor tile called NoScrub that simulated terrazzo tile and wood patterns; and improved materials for shoe soles and heels, led by Neolite Flex, Crown Neolite, Neolite Crepe, and Elasto Crepe.

*Top, U.S. Air Force Teracruzer truck and Translauncher semi-trailer, with TM-61B Matador missile, 1955. The eight-wheel-drive Teracruzer was built by GAC and equipped with the company's huge, high-flotation, low-pressure Terra-Tires. Bottom, cousin to the Terra-Tire, the Rolli-Tanker tires demonstrated their ability to move easily over rough terrain at Wingfoot Lake, 1956.*

Many transport and military airplane manufacturers approved a new lightweight aircraft brake, the Tri-Metallic disc brake developed by the aviation products division. Terra-Tires, which this division had developed in 1954 for Air Force missile support vehicles, were soon in demand for tractors, motorized golf carts, and mining equipment. They were cylindrical, barrellike, low-inflation tires that carried vehicles in a floating manner over rough or soft terrain.

In 1956, the Rolli-Tanker, an offshoot of the Terra-Tire, was developed for carrying fuel or other liquids. Similar to the Terra-Tire, the Rolli-Tanker is lined with a fuel-resistant synthetic and mounted on hub and axles; when filled, it can be driven easily over most terrain or floated in water.

Employment at Goodyear Aircraft Corp. was at 3,500 when North Korean military forces invaded South Korea in 1950, but topped 10,300 when the Korean truce was signed on July 26, 1953. During the rest of the '50s, GAC employment continued to be high — averaging about 8,500 — as military and nonmilitary equipment rolled out of the plants in Akron and Arizona.

Post-Korean War military equipment included such items as:

- Canopy assemblies and radomes for B-47E *Stratojet* bombers;
- Bondolite deck and bulkhead sections and other parts for F9F-9 *Tigers*;
- Booster cases for the Nike missile system;
- Metal radar platforms and antennae;
- Metal fuel tanks;
- Metal tow targets for jet gunnery practice;
- Electronic guidance systems for missiles;
- A line of analog computer equipment.

New products developed by GAC for the civilian market in the early '50s included plastic fertilizer hoppers, a plastic experimental pleasure boat 14½ feet long that weighed only 200 pounds, an inflatable airplane, made of rubber-coated nylon fabric, that could be packed in an automobile trunk when deflated, and a prototype one-man helicopter that weighed but 400 pounds and had a top speed of 60 knots.

After the Korean War, GAC's work evolved from the production of components, usually for airplanes, to the design and development of an entire rocket or missile system. As the company moved into aerospace production, it faced an increase in demand for research and advanced concepts in step with Space Age progress.

In 1955, a $3 million research and engineering building was added to the GAC complex in Akron, then a $300,000 engineering and laboratory building at Litchfield Park. By that time, GAC had become a major producer of booster metal parts for such military missiles as the Nike-Ajax, Nike-Hercules, Hawk, Matador, Mace, Farside, and Genie. In 1956, the Air Force signed a multimillion-dollar contract with GAC for Atran, an all-weather navigation system for manned aircraft and unmanned missiles. It was the result of 10 years of R & D efforts and was recognized as the most highly perfected all-weather navigation system yet produced.

In the late 1950s, GAC also supplied a wide range of missile support equipment for the Army and Navy and the Air Force's Strategic Air Command, including nose cones for the Army's Jupiter, the first successful U.S. Intermediate Range Ballistic Missile. On January 31, 1958, Jupiter put the first U.S. satellite in orbit.

## GAC: Looking to the Future

The most noteworthy of GAC's prime contracts in the missile field came in mid-1958 when the Navy gave it the job of developing Subroc, a complete antisubmarine missile system. The Navy's Bureau of Ordnance provided $65 million for this research and development assignment. Subroc would enable one submarine to detect another up to several miles away while both craft were submerged. It would then track and plot the target vessel's range, course, and speed and, when necessary, launch a missile toward the target.

At the end of the '50s, GAC had subcontracts with several leading rocket-missile manufacturers, including Boeing, Western Electric, Convair, and RCA, and it had become an established and recognized factor within the aerospace industry.

A peek into the future by Goodyear Aircraft engineers quickly and dramatically pushed Goodyear into the public spotlight as an integral member of the nation's Space Age industrial team.

At the eighth annual conference of the International Astronautical Federation in Barcelona, Spain, on October 9, 1957, Darrell C. Romick, head of the astronautics group of GAC's weapons system department, gave a forward-looking presentation on a three-stage ferry rocket system for establishing a manned earth satellite in space. He based his ideas on a 68-page summary of the system he had prepared with Richard E. Knight, manager of GAC's aerodynamics department, and Samuel Black, senior aerodynamist.

A week earlier, Goodyear had released advance news on the meeting in Spain and copies of the study with artist's illustrations of moon landings and interplanetary travel that the proposed system, called *Meteor Junior*, would provide.

News media were asked to hold the story until Sunday, October 6, but two days before the 6th, Russia launched *Sputnik I,* the first man-made satellite to orbit Earth. Newspeople deluged Goodyear with requests to publish the *Meteor Junior* material early, and they got immediate go-aheads. As a result, *Meteor Junior* produced the most newspaper publicity a Goodyear story had ever amassed until then.

## Blimp Endurance Records

As the world and Goodyear moved swiftly into the Space Age, the company's low-flying, slow-flying airships — the blimps — continued to perform for the military and as goodwill ambassadors for the company. GAC supplied 4 blimps to the Navy in 1954, and in May of that year a Navy Goodyear-built blimp set a new record for powered aircraft flight without refueling, remaining aloft for 200 hours and 4 minutes. Annual production of Navy blimps reached a post-World War II high of 23 in 1955. Two were equipped for Aircraft Early Warning service with radar housing set amidships atop the envelope, and 6 were a new type especially designed for antisubmarine operations.

In March 1957, a Navy blimp built by Goodyear established three flight endurance records in one trip. A ZPG-2W, about twice as large as the blimps Americans are now accustomed to, began a flight at the South Weymouth, Massachusetts, Naval Air Station, proceeded to Africa, then recrossed the Atlantic to terminate at Key West Naval Air Station in Florida. The journey lasted 11 days and 2 hours, covered 9,448 miles, and was the first nonstop round-trip flight over the Atlantic Ocean between the United States and Africa by a lighter-than-air craft.

Karl Arnstein, engineering vice-president for Goodyear Aircraft for 15 years, retired in February 1957. He was universally recognized as a leader in airship technology and in 1958 was awarded the Navy's Public Service Award, the highest honor the Navy then bestowed on civilians.

*Opposite, Subroc missile on submarine. Subroc, developed by GAC for the Navy in the late 1950s, enabled a submarine to detect another submerged craft several miles away and track and launch a missile toward it. Above, Navy blimps. Production of blimps for the Navy continued after war's end, with a postwar high of 23 in 1955.*

*A tapper at Goodyear's Dolok Merangir plantation on the Indonesian island of Sumatra deftly cuts a rubber tree for latex for use in tires and other rubber products. Dolok Merangir, site of Goodyear's first rubber plantation, was carved out of the jungle in 1916.*

# Chapter X

# GOODYEAR INTERNATIONAL

By Goodyear's 55th anniversary, in 1953, the employee register had stretched from the original 30 or so names in 1898 to 103,000. The corporation stepped up recruitment of college graduates that year for all production segments as personnel experts canvassed 157 campuses to sign up 500 seniors. The variety of manufacturing operations and the worldwide dispersal of the work force complicated employee relations, but management continued to stress the importance of people to progress and to nourish the time-honored ideas of Goodyear family and Goodyear spirit.

In many areas of industrial relations, company policies had advanced beyond the goals of labor relations established by the powerful labor movement of the late '30s and the New Deal. A new insurance program for all employees that provided life and nonoccupational accident coverage entirely at the company's expense was adopted in 1950. The amount of coverage related to individual earnings and increased as earnings rose. In 1953, an insurance plan that included hospitalization and surgical benefits went into effect. Overall, the medical benefits program has consistently improved since then, always ranking among the most comprehensive in U.S. industry.

Employee recreation, a vital force in the Goodyear family philosophy, grew with the company. An Employee Activities report published for the anniversary disclosed that the recreation program included 30 bowling leagues, 90 basketball teams, 87 softball teams, 30 golf leagues, and more than 60 clubs or social groups organized for employees or their families. The Goodyear Hunting and Fishing Club was the largest industrially sponsored club in the country.

In its 55th anniversary year, Goodyear also observed the 40th anniversary of its sponsorship of scouting. In 1913, it had been the first corporation to sponsor a Boy Scout troop and by 1953 was the world's largest industrial sponsor of Boy Scouts. Of its 36 troops, 11 were in Akron, 21 in plant cities in the United States, and 4 in plant cities abroad.

One of the largest displays of Goodyear spirit regularly occurred at Euclid Beach Park in Cleveland, scene of the annual company picnic for Akron-area employees and their families. About 50,000 people attended each year throughout the 1950s, even in '54 during the company's first nationwide strike, called by the United Rubber Workers on July 7 when the company and union were unable to agree on a new wage contract. Factories in 10 cities, including Akron, were affected, but after a 51-day shutdown, settlement was reached on August 27.

At the picnic that year, President Thomas placed the strike in its proper perspective against the backdrop of overall employee relations. "We have had these picnics year after year, and something more important than what is in progress now would have to occur to force cancellation of this traditional affair for our employees."

## Consolidating Non-U.S. Activity

On the eve of its 60th anniversary, Goodyear's foreign business had multiplied and extended far beyond the most hopeful dreams of 1910 when the company's first factory outside the United States was established in Bowmanville, Canada.

Foreign sales and profits made up nearly a third of the corporate total. Goodyear products were sold in every country outside the Communist bloc and promoted in several hundred languages. Employment outside the United States totaled 40,500, and factories were in 17 other countries. Managing the international operations became increasingly complicated as foreign subsidiaries and sales representatives abroad worked under many thousands of different national, regional, state, and local laws and regulations.

Management faced the need to consolidate the supervision and guidance of worldwide activities in such diverse and widely separated countries as Australia and Argentina, Sweden and Indonesia, France and South Africa, Luxembourg and Brazil. Accordingly, Goodyear International Corporation was set up in February 1957, replacing both the Goodyear Tire & Rubber Export Co., which had been responsible for foreign sales, and Goodyear Foreign Operations, Inc., which supervised foreign manufacturing. The new subsidiary, known immediately as GIC, was a management organization concerned with the policies and operations of business abroad.

GIC's first president was Frank Magennis, who had been vice-president and general manager of Goodyear Export and Goodyear Foreign Operations since 1954 and, for the 10 years preceding that, vice-president and assistant general manager of Export. E.E. Long, assistant general manager of Export, and George Hinshaw, vice-president and production manager of Goodyear Foreign Operations, were named VPs. All would celebrate 40th service anniversaries later in the year.

*Below, lower left, construction of Goodyear International's Tire Technical Center, Colmar-Berg, Luxembourg, 1969. GIC established the center in the Colmar-Berg tire factory in 1957 to develop and test tires for Europe. The center grew in scope, soon serving all of GIC's tire manufacturing operations, and was moved to new, modern quarters in 1969. Opposite, tree tapping at Goodyear's Sumatra plantation, 1957.*

Just as GIC came into being, three veterans who had played important roles in the company's overseas expansion retired. They joined about 1,200 others affected by new rules under an agreement with the URWA for compulsory retirement at age 65 for hourly employees and 68 for salaried personnel.

A. G. "Gus" Cameron had joined Goodyear in 1913, moved to export sales in 1919, and in 1922 became vice-president and general manager of Export at its founding. In 1939, he assumed similar responsibilities with Goodyear Foreign Operations, also at its founding, and headed both organizations until 1954 when Magennis succeeded him. He continued as a VP on the president's staff until retirement.

Walter A. Hazlett, who retired after 51 years of service, joined Export in 1937 as managing director of Goodyear-Great Britain and held that position 17 years before returning to the states as vice-president, special assignments.

J. J. Blandin, involved in crude rubber operations for all his 45 years with Goodyear, was an industry pioneer in growing and marketing crude rubber. While stationed in Singapore, he was sent to Sumatra in 1916 to conduct the initial surveys that led to the company's first purchase of rubber-growing land. He remained in the Sumatran jungle to oversee the task of carving out and setting up Goodyear's first rubber plantation, Dolok Merangir. He was vice-president of the Goodyear Rubber Plantations Co. and manager of the crude rubber division from 1925 until his retirement.

Although most of the international trailblazers were gone by 1957, the pioneering spirit remained strong among those who carried on, and until jet aircraft brought domestic and foreign operations within hours of each other, GIC was clothed in an aura of adventure and romance, which, in fact, it will probably never entirely lose.

## The Glamour Jobs

It was a rare GICer — as Goodyear International personnel became known in the company — who did not have exotic and exciting tales to tell on infrequent visits to Akron. Those tales were full of donkey rides to sell earth-mover tires in the Andean Mountains, business haggles with maharajahs, sacrifices of animals to consecrate new buildings in Southeast Asia, harrowing earthquakes in

Japan, and winter tire testing in the Swiss Alps. Lonely auto trips on the edge of the Sahara were among the stories; unbelievable bargains in the shops of Hong Kong; night glimpses of tigers in the Asian jungles; long, polite, but tough negotiations with distributors in the Mideast; and, almost always, Lucullan dining in Paris, Rome, Tokyo, Rio de Janeiro, Athens, Stockholm, or other world capitals.

No doubt many stay-at-home Goodyearites envied the romantic-sounding lives of their associates abroad, but whether the romance was real or imagined, tire making and marketing overseas were demanding jobs, conducted on a hardheaded, practical basis.

From the beginning, GIC operated on the theory of geographic profit centers. Each foreign subsidiary was a distinct operating function and profit center with its own profit-and-loss account and its own management team reporting to GIC headquarters in Akron.

In 1957, manufacturing-marketing subsidiaries existed in Argentina, Australia, Brazil, Canada, Colombia, Cuba, England, Indonesia, Luxembourg, Mexico, Peru, the Philippines, South Africa, Sweden, and Venezuela, and the British company had a second tire plant in Scotland. All but the Canadian company were under the wings of GIC. Sales subsidiaries were in France, Germany, Italy, Japan, Malaysia, New Zealand, Portugal, and Switzerland. Sales in countries with no subsidiary were made through distributors who supplied the retailers.

## Tropical Planters: A Breed Apart

With the increased production and marketing of products in foreign lands, Goodyear steadily developed a multinational character growing out of a perspective that had to be global. American GICers and their children enjoyed home life in circumstances usually different from those in Akron or other U.S. hometowns. That also held true for a rising number of high-potential non-American employees in assignments outside their native lands.

Perhaps the greatest change in life-style was experienced by those in plantation operations, where assignments brought them to the Goodyear rubber estates in the jungles of Brazil, Costa Rica, Guatemala, Indonesia, Panama, and the Philippines. W. E. "Moe" Klippert, who retired in 1967 as vice-president of rubber plantations and rubber purchasing, considered plantation people — tropical planters, as he called them — a separate breed.

In his autobiography, *Reflections of a Rubber Planter,* Klippert quoted a long-time associate in the rubber-planting industry to describe that special breed, the expatriate who manages, administers, and lives on a plantation. "To be happy and successful as a rubber planter," his friend had said, "one needs the mind of a philosopher, the habits of a hermit, and a good pickup truck."

A dedicated outdoorsman — and a good Dixieland jazz musician — Klippert was devoted to life and work in the plantations. He got his first plantation assignment in 1929, to Dolok Merangir, leaving there in 1932. He returned to Sumatra in 1934 on assignment to the Wingfoot Estate, established in 1928 about 160 miles from Dolok Merangir.

He describes this return in his book and provides insight into the character and attitude that make the ideal planter.

> I was in high spirits as we landed in Sumatra after an overnight voyage from Singapore on a small interisland steamer. To me it was coming home after a long absence, as I drank in the incredible beauty of the lush island. After a day or two at Dolok Merangir headquarters, we set out for Wingfoot, an all-day drive, the latter part of which alternated between newly opened estates and virgin jungle.
>
> Wingfoot Estate was the largest single rubber plantation in the world. Its 40,000 planted acres ranged from the oldest plantings, which were ready to come into production in 1934, down to the youngest area which had just been completed . . . permanent housing had been recently built on the estate, and we moved into a comfortable new house.
>
> Isolated as it was, Wingfoot Estate of necessity was largely self-contained. We operated our own sawmill, brick plant, machine shop, water system, electric system, ice plant, and contemporary motor road and light railway systems. We even made our own roof and floor tile. We operated a well-equipped hospital and provided housing, schools, mosques, provision shops, and recreational facilities for our labor force of workers and dependents, a total of about 10,000.

*A new planting of year-old rubber trees at Dolok Merangir.*

It was a vast and interesting operation, supervised by four Americans and some 20 Europeans, mostly Dutch. Most of our supplies came 60 miles up the Bila River from the small coastal port of Labuan, transported in a fleet of our own lighters and launches . . .

One evening shortly after we had become settled in our new home, I decided to show Anne [his wife] one of my favorite places on Wingfoot, a steep-sided, round-topped hill towering several hundred feet above the surrounding rubber, and called, because of its unusual shape, Gunung Tempurung (Mount Half-Coconut Shell). We stopped our car on the closest division road and made our way on foot through the young rubber to the base of the hill and then on a winding trail around the steep sides until we had reached the summit. From this point one could see right over the wall of virgin jungle that formed the nearby west boundary of the estate and continued unbroken to the mountain ranges miles away; and in the other direction there were rolling acres of Wingfoot rubber trees, stretching away to the horizon.

As we sat there shortly before twilight drinking in this tremendous view, I turned to Anne and in great animation said, "Wouldn't this be a wonderful place to build a house?" When she made no immediate reply I turned to look at her and found her gazing intently at me, on her face a curious mixture of bewilderment, wonder, and incredulity, and I realized suddenly and for the first time in my life that I must be different from most people. In time, of course, I came to understand and accept that good tropical planters are indeed a separate breed whose happiness in their outdoor world is neither shared nor understood by most other people.

In the decade following World War II, an annual average of about 50 tropical planters worked and lived on the seven Goodyear rubber estates, two each in Indonesia and the Philippines and one each in Brazil, Costa Rica, and Guatemala. They supervised the work of nearly 10,000 plantation employees, virtually all of whom lived on the estates with their families.

*A. J. Foyt in the winner's circle at the Indianapolis 500, 1977. Foyt's Indy 500 victory in 1967 had made him the first Goodyear driver to win at Indy in almost 50 years, helping to put the company on the road to dominance as supplier to top-flight motor racing.*

# Chapter XI

# 'GO, GO, GOODYEAR'

Top-management changes dominated the front pages of the *Wingfoot Clans* in 1958. In January, Victor Holt, Jr., was elected vice-president of sales, succeeding retiring Executive VP Robert S. Wilson, who for 30 years had headed the company's sales operations. Four months later, Sam Dupree, C. Colmery Gibson, Mark W. Laibe, and Oliver E. Miles were elected corporate VPs.

Dupree had been vice-president of the general products sales group and would continue to have this responsibility, also serving as executive liaison with all nontire divisions. Laibe, former director of purchases and assistant to the president, would head up all purchasing, general merchandise and materials control, traffic, warehousing, and the rubber plantations. Gibson would be responsible for the automotive products sales division and original equipment sales to automotive manufacturers. Miles had been sales manager of the tire division and VP of the replacement sales division; he would now be a corporate vice-president, replacement sales.

In October, E. J. Thomas was elected chairman of the board and Litchfield honorary chairman. Russell DeYoung, executive vice-president, succeeded Thomas as president, and P. E. H. Leroy, executive VP, was elected vice-chairman, continuing as the chief financial officer.

Dupree was elected vice-president of production and was succeeded as coordinator of the company's general managers by Richard A. Jay, formerly assistant to the president.

DeYoung, Goodyear's ninth president, was born in New Jersey, but came to Akron as a teenager. He earned a BS degree from Akron University and an MBA from MIT under a Sloan Fellowship. Like Thomas, he had overseas experience, serving in Java during construction and start-up of the tire plant at Bogor.

In 1940, he was named assistant to the president of Goodyear Aircraft and in 1943 was made vice-president in charge of production there. He was elected VP in charge of all Goodyear production in 1947 and executive VP in 1956.

## 'What Are You Waiting For?'

A fervent disciple of physical fitness, an energetic, hard-driving manager, instinctively a builder, DeYoung carried on the "get it done" tradition of the early days. He was a charger. His "What are you waiting for?" would become well-known to many Goodyear executives during his tenure as president and from 1964 to 1974 as board chairman. He seemed well-suited for the job of building on Goodyear's leadership and moving the company across new frontiers.

DeYoung had begun his rubber industry career with the company's closest competitor, Goodrich, but only briefly. As a freshman at Akron University he had wanted to join Goodyear, but was blocked by a company rule that allowed no college student to be accepted until completion of one year of study; so he went with Goodrich. He switched to Goodyear when he became a sophomore and spent the next four years building tires, manning curing equipment, and handling several other production jobs. He got his degree in industrial engineering, was assigned to the production squadron, and received the Litchfield Award of Merit as the outstanding member of his squadron's graduating class.

His first full-time Goodyear job was as a tire inspector, the beginning of an upward path through the production ranks that would influence him throughout his career; even in his days as chairman, he seemed at home on the factory floor. In frequent plant tours in the United States and abroad, he was greeted as Russ by many machine operators who had worked with him in the old days.

## Hands-On Leadership

Despite the organization's swelling size and complexity, Thomas and DeYoung continued to manage personally from the top. The company's brass, especially the chairman and president, remained accessible to all, as Seiberling and Litchfield had.

Admirers of Goodyear's personal management style pointed out that "nothing gets out of control in that company." The chances of any segment of it going sour without the top executives knowing about it were almost nil, they said. On the other hand, some critics claimed the top men were sometimes too involved in operations, inhibiting freedom of decisions and action at lower management levels.

The nature of the company, though, the training of its managers, their movement upward — almost always through the ranks — and the family spirit cultivated from the early days of Litchfield, made it difficult to imagine Goodyear's top executives either aloof from the action or managing the company by slide rule. In personality and corporate environment, they were people-oriented, and they obviously felt that effective human relations required good communications to and from top management, mixed with close personal supervision and control.

## Tribute to 'Litch'

On March 18, 1959 — less than six months after retiring as board chairman — P. W. Litchfield died in Phoenix following an operation. He was 83. Less than a month earlier he had attended a board meeting in Akron and appeared in good health. In a moving tribute to his long-time boss, associate, and friend, E. J. Thomas said:

> Mr. Litchfield was one of the truly great men of our times. The principles by which he lived and worked are firmly woven into our company and organization and will never die. His name and works will be remembered as long as there is a Goodyear Tire and Rubber Company.
>
> During the 58 years that he was actively associated with the company, he attained worldwide recognition as one of the great leaders of American industry and a pioneer whose outstanding contributions to the advancement of ground and air transportation were legion.

An *Akron Beacon-Journal* editorial of March 19 described Litchfield as "a teacher, a scientist, a philosopher, a salesman . . . and one of the great industrial leaders of the Twentieth Century." All obituaries listed his achievements in leading Goodyear to world preeminence, and many expressed the opinion that his greatest legacy was the Goodyear spirit of which he had been so proud and that he had nurtured throughout his long career.

Expansion of the company's foreign production capability, which had been one of Litchfield's constant aims from the 1920s on, gained considerable momentum in the year of his death. Construction was started on tire factories in Amiens, France, and Medicine Hat, Alberta, Canada, and plans were announced for a tire plant near New Delhi, India. Arrangements were completed for tire manufacturing in Italy by Ceat Rubber Co., in Chile by General Tire's subsidiary, and in Portugal by Mabor. Sales companies were established in Italy with Fred D. Aden as managing director and in Portugal with C. J. Alameda as general manager.

As Goodyear moved into the 1960s, its management widened the lead over competition. Expansion was a byword, and it was a rare plant manager in the states and managing director abroad who did not struggle for capital to increase production — and they got plenty of encouragement from the Akron hierarchy.

Appropriately, Goodyear's major U.S. tire advertising and promotion theme was "Go, Go, Goodyear." It headlined many print advertisements, was flashed on night signs of blimps — which by then had established themselves as bulwarks of the Goodyear image — and was featured in radio and television advertising. These dynamic three words characterized the sales and promotion campaigns, the hurry-up programs to build production capacity, and a corporate push for greater public recognition of the Goodyear name.

Go, Go, Goodyear dominated the '60s. With an eye on second-place Firestone at home and both Dunlop and onrushing Michelin in Europe, management pushed to extend Goodyear's sales lead in the United States, gain share in the growing European tire markets, continue leadership in Latin America, expand tire manufacturing in Asia, and enlarge the beachhead in Africa. These efforts and the dynamic growth of foreign tire sales were reflected in corporate gains during the decade: in sales that more than doubled from $1,550.9 million in 1960 to $3,215.3 million in '69, in net income from $71 million to $155 million.

In 1961, the company took over Geneva Metal Wheel Co. of Geneva, Ohio, a manufacturer of wheels for lawnmowers, boat trailers, farm implements, midget racing cars, and materials-handling vehicles.

Three years later, it acquired Motor Wheel Corp. of Lansing, Michigan, the world's largest manufacturer of styled wheels for autos. As a wholly owned subsidiary, Motor Wheel continued to produce wheels for autos, trucks, buses, and farm equipment; hubs and brake drums for cars and trucks; and air conditioners, heaters, and undercarriages for mobile homes.

The company had seven plants: three in Illinois — Chicago Ridge, Chicago, and Mendota; and one each in Lansing, Michigan, Newark, Delaware, LaGrange, Indiana, and Anaheim, California.

*Glow tires. Although never offered commercially, Goodyear's translucent tires, lighted from within by 18 tiny bulbs, typified the aggressive, dynamic promotions and "Go, Go, Goodyear" spirit of the 1960s.*

Motor Wheel Corp. had been founded on January 17, 1920, through an amalgamation of three Michigan companies — Prudden Wheel, Auto Wheel, and Grier Pressed Steel — and Weis & Lesh Manufacturing Co. of Memphis, Tennessee. The first president was Harry F. Harper of Prudden. Motor Wheel grew to become the largest U.S. producer of wooden-spoked wheels, which were then used by more than 90 percent of all vehicles on the nation's roads.

In 1922, it purchased the patents and assets of the Detroit Pressed Steel Co. and Forsyth Brothers Co. of Harvey, Illinois, producer of the Forsyth steel wheel. By 1924, Motor Wheel was the world's largest manufacturer of both wooden and steel wheels, producing more steel wheels for passenger cars than all other manufacturers combined.

Through its new Duo-Therm Division, Motor Wheel began manufacturing oil-fired water heaters in 1928 and soon added oil furnaces and oil space heaters. In the 1930s, it expanded product lines to include brake drums, wheels for agricultural implements, earth-moving equipment, mining and logging machinery, concrete mixers, power shovels, road rollers, and trailers.

During World War II, Motor Wheel shipped $155 million of ordnance materials to the armed forces. In 1953, it introduced the Electrofuse welding process and became the first company to manufacture welded passenger car wheels on a production basis. Motor Wheel acquired the Foreman Manufacturing Co. of Chicago in 1960, a producer of electric brakes and axle assemblies for mobile homes.

In 1964, the year of its acquisition by Goodyear, John H. Gerstenmaier, destined to become Goodyear's president and vice-chairman, became the fifth president of Motor Wheel.

New Kelly-Springfield tire plants were constructed at Tyler, Texas, in 1961 and Freeport, Illinois, in 1963. Two years later, Goodyear purchased an idle tire plant at Conshohocken, Pennsylvania, from Lee National Corp. and also the Lee tire business and retail outlets. Another wholly owned subsidiary, the Lee Tire & Rubber Co., was established to operate the plant and market tires under the Lee name.

*The world's largest tires in 1971, built at Goodyear's Topeka plant, were 11½ feet high and weighed 7,000 pounds.*

Lee had been founded in 1883 by J. Elwood Lee to make surgical supplies. In 1905, his company merged with Johnson & Johnson of New Brunswick, New Jersey, and five years later he started a new business, tire manufacturing. Lee died in 1914, and in 1962 Lee Rubber & Tire Corp. was taken over by New York interests and renamed Lee National Corp. The Conshohocken factory had been closed down by a strike from 1963 until Goodyear acquired it.

Goodyear's 10th tire plant in the United States — at Danville, Virginia — began production in 1966.

## Pushing More than Tires

The go, go attitude of the early '60s emphasized many products besides tires, and the company's capability in nontire lines increased dramatically in that period.

Construction of a $20 million synthetic rubber plant was started at Beaumont, Texas, in 1960 to produce Budene and Natsyn and to manufacture isoprene, the oil-derived chemical from which the Natsyn is made. Budene, Goodyear's polybutadiene, was a partial replacement or extender for natural or synthetic rubber that added durability and aging resistance. It was given the trade name Tufsyn in 1962 and used widely in tires, industrial rubber goods, and shoe products.

Chairman Thomas said that the Beaumont plant's planned annual production of 40,000 tons of Natsyn equaled the annual yield of 150,000 acres of rubber trees.

"Our ability to produce all the rubber we need within our own country will have two enormously important results," he said, "the near-elimination of the violent price fluctuations of natural rubber, which have plagued the rubber industry since its beginning, and complete independence of foreign sources of rubber for the United States and the rest of the free world, if necessary."

Freedom from dependence on foreign natural rubber came quickly. In 1954, U.S. rubber consumption had been split about equally between natural and synthetic, and the rest of the free world used about 90 percent natural and 10 percent synthetic. By 1959, the United States used about twice as much synthetic as natural and in non-Communist countries the ratio was 31 percent synthetic, 69 percent natural.

The foam products division got two new plants in the '60s. One, at Logan, Ohio, was established in 1963 to produce padded dashboards and other padded items for auto interiors. The foam parts were covered with Goodyear Neothane, a synthetic rubber that simulated leather and fabric textures with great fidelity and in any color. A factory in Bakersfield, California, was set up in 1964 to make urethane boat seats for West Coast manufacturers. In 1968, an automotive foam products plant was established at Luckey, Ohio.

A new shoe products plant at Madisonville, Kentucky, and an industrial products factory in Marysville, Ohio, went up in 1966. The Marysville plant concentrated on belting products.

## Olympic Gold Medals

From the company's earliest days, life at Goodyear for many meant more than just working on the job. The recreation program had spread to all factories by World War I and continued to grow through the years. A high percentage of employees participated, and it was a rare Goodyearite who could not find a company-sponsored or company-encouraged outlet to accommodate an active interest in sports.

By the 1960s, more than 40 employee recreational organizations conducted sports activities within the framework of programs at U.S. sites. In 1963, Goodyear received the Helms Foundation Award, presented annually to a U.S. or Canadian industrial manufacturer with the most outstanding year-round employee recreation program.

The sports program included bowling, golf, tennis, fishing, hunting, riflery, baseball, flying, model railroading, soccer, softball, volleyball, and cricket (in Great Britain), among other activities. The most visible sport, though, both inside and outside the company, was basketball. In the 1950s and '60s, Goodyear was an important force in amateur U.S. basketball, also on the international scene.

*1967-68 Akron Goodyears basketball team and coach Hank Vaughn (right). Over 60 years of varsity basketball produced company teams that won major championships at home and abroad, and players who captured five Olympic gold medals.*

Goodyear Hall's huge gymnasium no doubt was one reason for the company's adoption of basketball as its major "varsity" sport. Intra-company competition had been held there almost from the hall's opening day in 1920. Even before that, Fritz Page had coached the Goodyears to the 1918 championship of the American Industrial Athletic Association.

The main basketball court was constantly in use over the years by company leagues, industrial leagues, Midwest Conference competitors, and, in the 1930s, the tough National Basketball League, amateur forerunner of the National Basketball Association. In the '50s and '60s, the Goodyear Wingfoots — the varsity — got as much coverage in the *Wingfoot Clan* as any other single subject. In the 1960s, they were regularly atop U.S. noncollegiate basketball standings.

As Goodyear personnel experts annually canvassed 150 college campuses for seniors with management potential, both the employee recreation director and the director of athletics — who also served as basketball coach — kept watch for seniors with management potential *and* top ratings in college basketball.

Wingfoot squads in the 1950-70 period included several all-Americans and a host of all-conference stars. In 1961, C. E. "Chuck" Bloedorn, a member of Goodyear's National Basketball League championship team of 1937-38 and former coach of the Wingfoots and director of the employee recreation program, was appointed to the U.S. Olympic Basketball Committee. Two years later, Henry "Hank" Vaughn, the Wingfoot's coach since 1953, was appointed assistant coach of the Olympic basketball team.

In 1961, the Amateur Athletic Union (AAU) sponsored a tour of Italy, France, and the Mideast by the Goodyear team, which won all 22 games. While on a business trip to Turkey,

Chairman Thomas attended one of the games and learned of the Turkish government's interest in developing athletics in its national school system. He made possible Wingfoot Coach Vaughn's assignment to the Turkish Olympic Committee for three months to help develop a basketball program for high schools and colleges there. In 1962, the U.S. State Department sponsored another Wingfoot goodwill tour, this time to nine countries in the Far East.

The 50th anniversary of Goodyear varsity basketball, 1963-64, was perhaps the high point of the company's long association with the game. That season the Wingfoots won both the Midwest Industrial Basketball League and the National AAU championships, and three squad members — Larry Brown, Dick Davies, and Pete McCaffrey — gained places on the U.S. Olympic team, winning gold medals at Tokyo when the United States defeated the USSR team for the championship. Hank Vaughn was assistant coach of the U.S. squad.

During 1965-66, the Goodyear team won the championships of the National Amateur Basketball League, the AAU, the prestigious Finland Cup competition at Helsinki, and, in January 1967, the Intercontinental World Cup tournament at Rome, beating teams from Europe, North America, and South America.

Two Goodyearites were named to the 1966 U.S. team that won at the Pan American games, with Vaughn as head coach. Goodyear stars Calvin Fowler and Jim King were members of the U.S. squad that won the Olympic title in 1968 at Mexico City, where it defeated Yugoslavia in the final game. The Goodyears won their second Intercontinental World Cup title in 1968, defeating Spain at Philadelphia, and its third in 1969 at Macon, Georgia, when it beat Spartak of Czechoslovakia. The squad made short playing tours of Mexico in 1968 and Brazil and Venezuela in 1969.

In April 1970, at the height of an industry-wide strike by the United Rubber Workers, the company "reluctantly discontinued" its varsity basketball program. This move was part of an effort to reduce expenses in the face of the costly strike. Other reasons were the previous withdrawal of many major manufacturers from big-time industrial basketball competition and the attraction of professional basketball for many college graduates who, in years past, had played as amateurs during careers in industry.

During the half century and more of varsity basketball, from 1914 to 1979, about 300 employees played for the Goodyears, or Wingfoots, among them a board chairman, E. J. Thomas, on the 1920-21 squad and a corporate president, Victor Holt, Jr., on the 1929-30 team.

## Return to the Speedway

In Goodyear's first two decades, its management had capitalized on motor racing as a promotion vehicle and tire development tool. As early as 1902, Litchfield had tried out some of the early straight-side tires in English races. Just before the U.S. entry into World War I, top U.S. drivers established several world records while using the improved straight-side cords. In the first Indianapolis 500 race after the war, straight-sides rolled to first and second places and were on all but one of the 14 finishers. That year, however, Goodyear ended its involvement in racing because, it said, of decreased value in tire development and design. Goodyear had, of course, made its point and proved its expertise in tire making.

The company returned to racing in the mid-'50s, quietly at first, but then with a roar. By late in the next decade, it was the dominant supplier to the major forms of auto racing in the United States. By the early 1970s, it held the same position in Grand Prix racing, the most prestigious and visible of international motor sports.

Automobile racing boomed as a spectator sport in the 1950s. More than 200,000 fans jammed the Indianapolis Speedway each Memorial Day for the classic and world-famous 500-mile race of races. Attendance at Grand Prix events around the world averaged about 100,000 people per race. Marathon races for sports cars — such as the 24 Hours of Le Mans and the 12 Hours of Sebring — attracted even larger crowds. Drag racing became highly popular in the United States, and TV began to cover all big racing events on a regular basis.

An even more important incentive — and perhaps a somewhat aggravating one — for the world's largest tire maker to reenter racing was the promotion of success in racing by the two dominant racing tire suppliers: Firestone in the United States and Britain's Dunlop in Europe. Firestone was the sole supplier to cars in the Indianapolis 500 and pegged much of its advertising on repeated victories in this famous race, which had become an American institution. Dunlop, the major tire supplier to the Grand Prix races, also made good use of racing as a promotional vehicle. Goodyear knew that a return to racing would be expensive, but other things had to be considered.

Opinion surveys in the mid-1950s disclosed that Goodyear had a strong appeal to the middle-aged and older generations and that its tires had a solid reputation for quality and reliability. This image had a trace of stodginess in it, though, and the company's attraction to the fast-increasing population of school-age Americans and those in their 20s was less than for older citizens. Management therefore saw the need for a more dynamic corporate aura. Hence the adaption of Go, Go, Goodyear advertising and promotion campaigns — and the return to motor racing after a 37-year hiatus.

The program began with a full line of racing tires for stock cars in 1958. Success came quickly. The winner and 7 more of the top 10 finishers in the Southern Stock Car race on Labor Day 1959 at Darlington, South Carolina, were on Goodyear Blue Streak tires. A few months later, in February 1960, the winner and 8 other first 10 placers in the Daytona 500, considered the most important race on the stock-car circuit, drove on Goodyears.

Victories followed in such high-visibility events as the Dixie 400, National 500, Trenton 100-mile Championship, Nashville 500, Atlanta 500, Daytona Firecracker 400, and Pike's Peak Hill Climb. Included in the list of top drivers winning big races on Goodyear tires in the early 1960s were A. J. Foyt, Dan Gurney, Gordon Johncock, Richard Petty, Johnny Rutherford, Carroll Shelby, LeRoy Yarborough, and other drivers of similar renown.

In NASCAR's (National Association of Stock Car Auto Racing) Grand National stock car season of 1962, Goodyear led all tire manufacturers in number of victories, percentage of finishers in the top 10, and number of drivers using its tires.

By 1964, Goodyear was almost ready to challenge Firestone's dominance of America's best-known car race, the Indianapolis 500. Tires were shipped to the famous track, but none was used in the qualification trials; the company's racing division, which had been established in February 1959, had to wait another year.

Loaded with Go, Go, Goodyear spirit and encouraged by a highly competitive management, in the spring of 1965 the tire engineers again invaded Indianapolis, but they were once more disappointed.

Twelve of the 33 qualifiers were on Goodyears, including Foyt, who had won the pole with an average speed of more than 161 mph, but he and 9 other Goodyear qualifiers had mechanical problems and failed to go the distance. Top finishers on Goodyear tires were eighth and ninth.

The score improved in 1966 when 16 qualifiers rode on Goodyears and Johncock came in fourth. Most Goodyear-shod entries had been eliminated, either because of a 16-car pileup on the first lap or mechanical malfunctions.

## Outclassing the Competition

The big payoff came in 1967 when the redoubtable A. J. Foyt powered his way to victory at Indy, the first Goodyear driver to do so since Howard Wilcox in 1919. Goodyear's dominance as the major supplier to top-flight motor racing in the United States was just a few years away.

Goodyear International's auto racing division had been making an auspicious start of its own in Formula One Grand Prix racing and other major phases of this increasingly popular sport.

Following the parent's example, International established a racing division in 1964 to direct the tire supply effort for racing outside the United States, particularly for Formula I Grand Prix events, Formula II racing, and such long-distance contests as the 24 Hours of Le Mans, usually attended by more than 300,000 enthusiasts. Its reasons were akin to those motivating renewed support in the United States.

In GIC's worldwide *Orbit* magazine, International's president, Richard V. Thomas, explained:

> Racing is a useful tool in our business kit. We go to school, so to speak, on race circuits from Le Mans to Singapore. We consider our racing activities — our experimenting on compounds in the laboratory, our innovating on design in the factory, our constant checking during testing and trial runs and in the heat of competition — as a very important part of our research and development program.

*Opposite, Jack Brabham (left) brought Goodyear its first Grand Prix world driving championship in 1966, and Denis Hulme (center) of New Zealand duplicated Brabham's feat the following year. Dan Gurney (right) teamed with A. J. Foyt in 1967 to win the 24 Hours of Le Mans in a Ford Mark IV on Goodyear tires for the first all American victory in the French classic. Below, Denis Hulme captures the Grand Prix world driving championship on Goodyear tires, 1967.*

The support of motor racing in Europe, where, except for Sweden and the Benelux countries, Goodyear was little known, also would help build corporate identification among car-conscious Europeans. Compared with Britain's Dunlop, France's Michelin, Italy's Pirelli and Ceat, Germany's Continental, and even some of the smaller tire makers, Goodyear was not well-known on the Continent. With a growing investment in Europe, boosted by new tire factories in France and Italy and with a German plant on the drawing board, the company needed greater European visibility and acceptance — and quickly.

The first major Goodyear victory in international racing came at the tag end of the 1965 Grand Prix season when American Richie Ginther, driving a Honda, won the Mexican Grand Prix. Firestone and Dunlop were still the leading suppliers for the major international circuit, but Goodyear's drive to the front accelerated a year later when Ford Mark IIs, on Goodyears, finished one-two-three in the 24 Hours of Le Mans, and Australian Jack Brabham won the Grand Prix world driving championship in his Goodyear-shod Brabham-Repco. Within seven years both Dunlop and Firestone would withdraw from major Grand Prix tire support.

## Land-Speed Records

Dominance as tire supplier to major U.S. and international motor racing came to Goodyear suddenly — even if not fast enough for its racing-tire engineers. But leadership was almost instant in the more exotic and much less crowded field of land-speed competition.

In 1960, a year after establishment of the racing division, Mickey Thompson of California equipped his four-engine, wheel-driven Challenger I vehicle with Goodyears and clocked a record 406.6 mph over a mile at Utah's Bonneville Salt Flats. That was only for the first leg of the two miles required for official recognition, however, and his engine broke down on the return mile. This denied him the world land-speed record, which remained at 394.2 mph, set by John Cobb of England.

The record fell on August 5, 1963, when Craig Breedlove of California and his *Spirit of America* on Goodyear tires posted an official average of 407.45 for Bonneville's measured mile. In the following autumn, the record went to Tom Green of Illinois, driving the jet-powered *Wingfoot Express* built by Akronite Walt Arfons. Green averaged 413.2 mph on Goodyear tires.

Breedlove came back just 11 days later to record 468.72 and on October 15 became the first man to drive at a speed in excess of 500 mph, averaging 526.28 for the mile-and-return course on the Bonneville Flats. The record then passed to Art Arfons, Walt's brother, on October 27 as he posted an even 536 mph at Bonneville on Firestone tires.

*Above, Craig Breedlove's "Spirit of America," 1965. Opposite, Breedlove (left) and Goodyear President Russell DeYoung, with model of "Spirit," 1963. Breedlove set land-speed records each year from 1963 through 1965 in his jet-powered vehicle, riding each time on Goodyear tires.*

A year later, on November 7, 1964, Art moved his record up to 576.553 mph. Two weeks later, Breedlove, in a new *Spirit of America* with wheels, brakes, deceleration system, and tires provided by Goodyear, became the first man to drive on land at more than 600 mph. He averaged 600.601 mph for the two-way run.

In three years — from 1963 through 1965 — Craig Breedlove had become the first person to crack the land-speed barriers of 400, 500, and 600 mph, each time on Goodyear tires.

The tremendously successful reentry into motor racing was just one element of Goodyear's corporate drive to strengthen its reputation as a progressive, dynamic, now-oriented company. In the mid-1960s, advertising and promotion programs keyed to the Go, Go, Goodyear theme and using such pioneering techniques as the distribution of millions of Goodyear Christmas long-playing record albums developed into the most aggressive in the rubber industry. Goodyear's blimp was fast becoming one of the best-known corporate symbols in the world.

Many insiders felt that the company's resurgence as a forceful, omnipresent, and persuasive communicator began to take shape in the late '50s after Chairman Thomas and President DeYoung decided to inject fresh blood into corporate communications. New managers were brought in from outside the company to succeed retiring veterans, and a director of business planning and research was appointed.

In 1958, Robert H. Lane was hired as director of public relations. He came from Carl Byoir & Associates, then the world's largest PR firm, where he was a vice-president. John P. Kelley, president of a leading Columbus, Ohio, advertising agency, joined Goodyear in 1959 and became advertising director in 1961. Edwin H. Sonnecken, president of Market Planning Corp. of New York City, arrived in 1961 and a year later was put in charge of the business planning and research program. Eventually all three men were elected as the company's first corporate vice-presidents in their fields.

The newcomers built one of the hardest-hitting communications programs in U.S. industry, perhaps most dramatically symbolized by the constant exposure of Goodyear blimps — both in the sky and the home — to most of the population of the United States.

## Ambassador in the Sky

Thomas and DeYoung gave full credit to Lane for the blimp's quick acceptance as a household word and sight. By fashioning the airship into a spectacular "night sign in the sky," he had greatly multiplied its value as an aerial ambassador, and he used the proliferating magic of television to fly the Goodyear name and good image into American homes by the multimillions.

When Lane first arrived in Akron from New York City with a management mandate to expand and improve Goodyear's PR goals, he was shocked by an apparent disregard of the public relations potential in the company's one remaining blimp, *Mayflower*. Based in Miami, *Mayflower* had become part of the local scene there, operating almost totally in southeast Florida. It carried passengers and sold commercial advertising time on its night sign.

Top management seemed convinced of the blimp's value, but had no clear ideas on how to take full advantage of it. Some company executives, rumor had it, thought the airship should be deflated and "put in a box" until somebody devised ways to make it pay for operating and upkeep expenses. Late in 1958, management actually was ready to take *Mayflower* out of service.

Lane made a plea to Thomas and DeYoung, then to the board of directors, to keep that one remaining blimp in operation for one more year. He requested full responsibility for programming *Mayflower*'s operations, and he got it.

In a six-month tour that followed, the airship traveled in a detailed plan north along the Eastern Seaboard as the first blimp tour solely from a public relations perspective. A full-time PR professional assigned to the airship also was a first. That was the start of a continuing campaign that soon made "Goodyear blimp" bywords throughout the United States and, 15 years later, in Europe as well.

The success of *Mayflower*'s 1959 junket buttressed Lane's conviction that the airship was a unique PR vehicle of tremendous potential. Top management — especially Thomas and DeYoung — soon shared his conviction. For the eager and aggressive public relations director, though, one blimp was not enough, particularly with his plans to expose the Goodyear name to millions of Americans over and over again through television. With support from Thomas and DeYoung, he made a presentation to the board requesting another blimp. In 1963 *Columbia* joined *Mayflower*.

Five years later, *America* became the third member of the fleet. Because in 1962 the Navy had scrapped its 17 remaining blimps — their early-warning and antisubmarine functions had been taken over by advanced radar, space satellites, and other aerospace developments — the only blimps still operating in the United States were those in the small Goodyear fleet.

Any doubt about the effectiveness and value of the blimps almost certainly had dissipated by the mid-'60s. *Mayflower, Columbia,* and *America* were then the most easily identified flying objects in the skies and were accepted scenery at some of the most popular televised events on the national sports and entertainment calendars.

The Goodyear blimps were in great demand by the news media, particularly TV, for use as aerial camera platforms. They were seen and mentioned regularly on national television as the networks used them to cover such spectaculars as football bowl games, the Indy 500, the America's Cup yacht races, the world series, major golf tournaments, and world's fairs. They even surveyed the migration of whales along the California coastline.

Each airship annually carried about 8,000 passengers, including celebrities, politicians, customers, journalists, and Goodyear dealers. Even former President Dwight D. Eisenhower was once an enthusiastic passenger. Using lighted, computerized signs at night, the blimps celebrated municipal anniversaries throughout the nation and supported public service and charitable activities and major Goodyear promotions. About three of every four of the night messages were for public service.

Surveys in the 1960s showed the blimps making millions of new friends, and — as Lane pointed out in his frequent pitches to management — "Nobody else has a blimp. It's like having the only ad in *Life* magazine, the only sign in Times Square, or the only billboard along the Los Angeles Freeway."

## 'Easily Identified Flying Objects'

The sight of a blimp floating in the sky or on a television screen undoubtedly evoked many different responses among the millions of Americans for whom the rotund nomads had become familiar. The adjective used most frequently to describe the blimp: friendly.

One of the most poetic descriptions of the nature of these sky creatures is in George Larson's text for *The Blimp Book.*

> Blimps exist today primarily to make people happy, and few other mechanisms are as successful in the execution of their purpose. But blimps evoke more than mere pleasure. These great tumid vessels have a unique allure. They induce a wild affection, and people who fall under their spell find themselves wishing somehow to embrace them.

*Opposite, a close-up view from below of a blimp gondola and the thousands of bulbs that make up the Super Skytacular night sign. Left, spectators move in for a close encounter with a blimp. Blimps draw enthusiastic crowds wherever they appear.*

> Blimps have moods. For the most part they seem to carry themselves with a certain gentle humor tinged with melancholy. They wander like so many grey eternal ghosts. They rise above the gloom at dusk to glow in the sun's fire. At night, they loom in opaque skies, like unseen phantoms bearing all our fears. Attracting the most innocent of children and the weariest of grownups, they promise in their shadows relief from care and trouble. A blimp is pure fun, devoid of evil purpose. It has no natural adversary. It is an anachronism, stretching out its glory, playing a role that has no conceivable end while the riddle of its allure remains unanswered. Its future can be measured not by the life expectancy of its moving parts but by the need to have it around, and that may last forever. Nobody doesn't love a blimp.

Goodyear's "easily identified flying objects" may have assumed a slightly sharper, more angular appearance when they took on the "Skytacular" night signs, which also increased the value of blimps as communications and public relations implements. Airships had first become message bearers when a Goodyear model of the 1920s was emblazoned with company name and trademark, painted on its large sides. Next came blimps trailing banners to advertise consumer products of Goodyear and other manufacturers.

*Top, the Super Skytacular night sign is capable of an infinite variety of four-color electronic designs. Bottom, technician at Akron's GAC lab creates airship message on CRT screen, using light-beam pencil.*

## Lights, Color, Action

The first installment of illuminated signs was in the late '30s. Lightweight frames carrying white neon tubing were strapped to the sides of the blimps. The tubing permitted formation of any number or letter within a frame, and each sign had 10 frames. After World War II, Goodyear engineers developed a sign with incandescent lamps instead of neon. It had 10 panels with 182 lamps on each, and an individual panel could form any letter or number. The same principle is used in today's pocket calculators and digital clocks.

In the late 1940s, one airship carried an illuminated sign that showed running copy, but no color or animation — an aerial reproduction of the familiar Trans-Lux sign in New York's Times Square. The airship sign's weight and the equipment necessary to run it required a large blimp; so Goodyear purchased a 250-foot K-type blimp, 90 feet longer than the *Mayflower*, from the Navy. The high cost of operating it was a factor in the decision to discontinue use of the Trans-Lux sign.

The Skytacular sign, which uses both color and animation, was developed by Goodyear Aerospace in 1964-65 and first installed on the *Mayflower* in 1966. This sign was conceived in Belgium in 1960 when Lane saw one like it operating in the famed Grand Place of Brussels.

"My God," he reacted. "Let's put that on the blimp."

Lane enlisted electronic engineers at Aerospace to develop a computer-operated, full-color animated sign that could be carried on each side of a blimp. The first Skytacular was 105 feet long and 14.5 feet high; each of the two on the airship's sides had 1,540 light bulbs.

Its first formal display was at the Wingfoot Lake facility — to an audience of one, Russell DeYoung, then board chairman. He had received a "rather mysterious" request, as he put it, from Lane to go there and "see something new." DeYoung said he could come out almost any day that week, but Lane said, "No. It has to be at night."

So the chairman drove out at night and was greeted by his public relations director and the Skytacular. He bought it on the spot and later helped Lane sell the idea to the directors.

A Super Skytacular, larger and improved, was introduced in 1969: 195 feet long, 24.5 feet high, with 3,780 lamps (actually auto taillight bulbs) on each side connected by 80 miles of wiring. On a blimp flying 1,000 feet above the ground, the supersign can be read from the ground a mile distant on either side. Messages for the sign were created on exotic electronic equipment in a Goodyear Aerospace lab at Akron with a technique, gradually improved, that is still used.

A technician draws animation and copy on a cathode ray tube (CRT) with a light-beam pencil, and from there a computer takes over. The process results in a magnetic tape. A six-minute tape consists of about 40 million bits of on-off information running through electronic readers aboard the blimp that control the bulbs, color selections, and speeds at which the messages move.

The magnetic tapes are being replaced by "floppy disks" which store much more information and give the blimps immediate access to hundreds of animated messages programmed in Akron.

Blimp value as a goodwill ambassador was dramatically illustrated during the 1965 water shortage in the Mid-Atlantic states, especially in metropolitan New York. After giving the city's Mayor Robert Wagner a flight over depleted municipal reservoirs in upstate New York, down to 49 percent of normal levels, *Mayflower* was joined by *Columbia* in a public service campaign urging New Yorkers to conserve water.

The two ships flew special night-sign messages, including, "Join New York's fight to save water," and "You, too, can save 25 gallons of water every day."

New York's six major daily newspapers and local radio and TV stations covered the campaign. Associated Press and United Press International carried pictures and stories of the blimps to newspapers throughout the country. When the crisis had ended, Mayor Wagner presented a certificate of thanks to Chairman DeYoung and praised Goodyear for "acting in the interests of the people of New York . . . in an outstanding example of generosity in public service."

Through their exposure in the skies, in newspaper and magazine stories, in feature motion pictures, and especially on TV, the Goodyear blimps had become part of midcentury Americana by the late 1960s. But few people — even most Goodyearites — were aware of the mechanics and logistics that with some refinements prevail even today.

## How a Blimp Operates

Each modern blimp is staffed with an air and ground crew of 23: 5 pilots, a PR representative, and 17 crewmen, including engine, radio, electronic, and structural experts. With four specially equipped ground-support vehicles, the crew is almost self-sustaining in the field in regard to operation and maintenance. Moving from city to city, the caravan travels by highway as the blimp flies to its next engagement.

A custom-designed bus serves as a flight center and communications headquarters. It is equipped with all the administrative aids necessary for operation and a special mast for landing in an emergency.

A tractor-trailer rig serves as a mobile maintenance facility and is equipped with a machine shop and a night-sign and TV equipment lab. The trailer also carries the main mast, spare parts, and supplementary equipment. A passenger van and a sedan round out the rolling stock and are used for ground liaison work and crew transportation. All vehicles have two-way radios for communication with one another and the blimp.

Each blimp can carry six passengers and a pilot. Its rubber-coated polyester fabric envelope is filled with helium, an inert, nonflammable gas. The blimps have a range of about 500 miles, a cruising speed of 35 mph, and a maximum speed of 53 mph. On a calm day, they can remain almost motionless, making them ideal flying platforms for TV and still cameras. They can fly at altitudes of up to 8,500 feet, but normally cruise from 1,000 to 3,000 feet off the ground.

Fully equipped with the most modern flight equipment, the blimps normally do not fly in adverse weather, and in more than a half century of operations, no peacetime passenger of a Goodyear blimp has ever been injured.

*Scenes of exotic contrasts like this native and his camel strolling past Goodyear's modern tire factory in Morocco became increasingly common in the 1960s and '70s as the company accelerated its foreign expansion, particularly in the Third World.*

# Chapter XII

# A MULTINATIONAL GIANT

As they climbed from the wreckage of World War II — with much help from the United States — the developed nations created automotive industries with voracious appetites for tires, and the developing nations were stepping into the Auto Age. Car production in the Western World, excluding the United States, quadrupled in the decade of the '50s, and truck production more than doubled. U.S. auto and truck output in that period was flat, with practically no gain, but burgeoning foreign markets beckoned to the U.S. tire industry, which included the four largest tire manufacturers in the world. It responded, and Goodyear was in the lead.

For many years, Goodyear had held a dominant position in Latin American tire markets; now it looked hard at Europe and Asia. Its market position in the European region was the weaker of the two. Until the late 1950s, Goodyear had just two factories on the Continent — in Luxembourg and Sweden — and two in Great Britain. It had no production facilities in West Germany, France, and Italy, by then the second, third, and fourth largest auto markets in the world.

The big push into Europe began in 1959 with the construction of a tire plant in France, at Amiens. This factory represented an initial investment of about $10 million and was the first U.S. tire manufacturing installation in that country. It went on stream in 1960, while across the Atlantic another tire plant began production, at Medicine Hat, Alberta, Canada.

Expansion in Europe continued full force throughout the '60s. In 1962, Goodyear acquired Gummiwerke Fulda, a leading German tire manufacturer. The headquarters and factory of the 62-year-old company were in Fulda, Hesse, about 20 miles from the Iron Curtain.

The company now had solid manufacturing footholds in two of Europe's Big Three tire markets, and Italy, the third, was next. An incursion into Italy had started in 1959 with the establishment of a sales company, Goodyear Italiana SpA, to market Goodyear tires manufactured by Ceat, the second largest Italian tire maker; Pirelli was the largest.

In late 1963, Goodyear took advantage of the Italian government's program to encourage manufacturing in the underdeveloped Mezzagiorno region of southern Italy and began building a tire and tube plant at Cisterna di Latina, 30 miles south of Rome. On May 6, 1965, the new $9 million factory was inaugurated. Chairman DeYoung and Goodyear International's president, Richard Thomas, were joined in the dedication ceremonies by Italy's minister of defense, Giulio Andreotti, who would be that country's prime minister in the mid-1970s, and Eugene Cardinal Tisserant, dean of the Roman Catholic church's college of cardinals.

The presence of national leaders at factory inaugurations outside North America had become a Goodyear tradition. To dedicate a new plant without a president, prime minister, premier, governor, or some other top-ranking government official on hand was just not done, especially as the public relations people hoped for widespread news coverage of the event.

Although the acquisition of Fulda in 1961 had increased corporate income from Germany, it had no effect on the Goodyear-brand share of the West German market; this plant made only Fulda tires. That situation changed in 1967 when a $15 million Goodyear tire factory at Philippsburg, 15 miles from the university city of Heidelberg, was completed. Deutsche Goodyear GmbH, which had been established in 1955 to sell Goodyear-brand tires made by Continental Gummiwerke, Germany's biggest tire manufacturer, would operate this facility. Philippsburg had the largest initial production capacity of any Goodyear plant built outside the United States, 2,400 tires a day. At long last, Goodyear tires were manufactured in France, Italy, and Germany, as well as the United Kingdom, Luxembourg, and Sweden.

Those tires differed a lot, though, from tires produced in the United States where, until the late 1950s, practically all of Goodyear's tire research, development, and testing had been carried out. The smaller cars, narrow roads, and high speed limits of Europe brought new and different technological requirements to U.S. tire engineers, who had to spend much time on the Continent studying the several dissimilar markets there and their peculiar requirements.

## R & D in Europe

In 1957, Goodyear International established the Goodyear Technical Center-Europe at Colmar-Berg, Luxembourg, alongside its six-year-old tire factory there. Its original purpose was to design, develop, and test passenger, truck, and off-road tires for Europe, working both with Goodyear factories on the Continent and in Great Britain and with U.S. research and development forces. This center also was a training ground for European technicians who eventually would take over its control from the U.S. technicians of the original staff.

*Opposite, Goodyear President Russell DeYoung (left) and future Italian Prime Minister Giulio Andreotti (second from right) look on as Cardinal Tisserant gives the blessing at inauguration ceremonies for Goodyear's new plant in Cisterna. Above, a rubber inspector examines smoke sheet rubber at the Goodyear Orient Company warehouse in Singapore.*

At the start, the center's technical staff numbered 3. By the late 1960s, it had increased to nearly 300 people of a variety of nationalities. The center grew rapidly in size and scope and soon served the entire spectrum of Goodyear International's tire manufacturing operations. In 1969, its name was changed to the Goodyear International Tire Technical Center.

As management moved boldly into the markets of industrialized Europe, it did not disregard the developing areas of the world. A tire and tube factory was completed in 1961 at Ballabgarh, India, on the outskirts of New Delhi. A Turkish factory at Izmit began producing tires and tubes in late 1962.

In North America, Goodyear-Canada acquired the Seiberling Rubber Company of Canada, Ltd., and its factory in Toronto in 1964. A year later, a new tire plant at Valleyfield, Quebec, began operations, along with a second tire factory in Australia, at Thomastown, near Melbourne.

## Bogor Plant Lost

Until 1960, Goodyear had lost only one factory through intervention of a foreign government, the tire plant at Bogor, Indonesia, which had been taken over by the Japanese army in 1941 and regained by Goodyear in 1946, following the Allied victory in World War II. Then in 1960, the Castro regime expropriated the 14-year-old tire factory at San Jose on the outskirts of Havana, Cuba. The third factory loss came in 1965 in Indonesia. For the second time, the Bogor factory and the plantation at Dolok Merangir passed from Goodyear control.

Following Indonesia's confrontation with Malaysia, the pro-Communist Sukarno government seized the two installations and several other U.S.-owned operations. These moves were in apparent retaliation for the United States' support of Malaysia. After the abortive Communist coup in September 1965 and the establishment of President Suharto's administration, however, both facilities were returned to Goodyear, which resumed control in 1967.

The Caribbean foothold lost in Castro's takeover of the Cuban factory was regained in 1967 through construction of a plant at Morant Bay, Jamaica. Goodyear acquired its first Central American manufacturing operation in 1968 with the purchase of a majority interest in Gran Industria de Neumaticos Centroamerican, a tire company with headquarters and a factory in Guatemala City. At that time, Firestone operated the only other tire factory in Central America, in Costa Rica.

As new industrial technologies created new, fast-growing markets for industrial rubber products and chemicals, Goodyear responded with increased exports from the United States and the establishment of several nontire plants overseas. A chemical plant to manufacture synthetic rubber for the European market was completed in 1963 at Le Havre, France, and in 1968 a new $16 million industrial products factory was opened in Northern Ireland at Craigavon, 20 miles from Belfast.

Capacity for industrial-products manufacturing in Canada also greatly expanded in the '60s. A new industrial hose plant opened at Collingwood, Ontario, in 1967, a year after a new foam products factory began operations at Owen Sound, Ontario. In 1966, Motor Wheel Corp. of Canada established a plant at Chatham, Ontario, to make auto wheels and tubular parts.

## Up Through the Ranks

On April 6, 1964, 48 years after joining Goodyear as a part-time clerk at $25 a month, Edwin J. Thomas retired as chairman of the board and chief executive officer and was succeeded by Russell DeYoung. Victor Holt, Jr., succeeded DeYoung as president.

Thomas, who for 24 years had been president or chairman and CEO, was requested by the directors to continue on the board and became chairman of the board's executive and finance committee. DeYoung, a Goodyearite for 36 years, had been president since 1958.

The towering, affable Holt had moved up through the sales ranks, following a different route to the presidency than his two predecessors, both production men. He started with Goodyear in 1929 as a sales trainee after graduation from the University of Oklahoma, where he was an all-American basketball standout. He held sales and sales managerial positions in various parts of the country before being named assistant manager of tire sales in 1943 and manager a year later. He was elected a corporate vice-president in charge of replacement tire sales in 1956 and executive vice-president in 1958.

Howard L. Hyde, who had been executive vice-president for financial and legal affairs since 1960 and a director since 1940, was named vice-chairman of the board, while continuing as chief financial officer.

Other major moves included the election of three new executive vice-presidents: Sam Dupree, who had been vice-president for production for six years; Oliver E. Miles, who had served as sales vice-president for six years; and Richard V. Thomas, who since 1961 had been president of Goodyear International. Thomas also continued in his GIC position.

Four new vice-presidents were elected, all with extensive service in their areas of responsibility: Albert J. Gracia, research; Richard A. Jay, general products; Walter H. Rudder, production; Charles A. Eaves, replacement tire sales. Two changes in the management of financial operations also were announced. Bruce M. Robertson, formerly vice-president and comptroller, was elected vice-president of finance, and J. Robert Hicks, assistant comptroller, was named comptroller. Both were relative newcomers, joining Goodyear after lengthy service in General Electric's finance operations. Robertson had joined Goodyear in 1961 as vice-president and comptroller, and Hicks had come on board in 1962 as assistant comptroller.

## Revolution in Transportation

The winds of change that would sweep the world were at magnum force in the mid-1960s. Europe and Japan had regained full industrial strength, and the scars of World War II had disappeared from both their landscapes and their psyches. The U.S. government and its citizens were developing stronger feelings of responsibility for world progress, peace, and the protection of interests abroad. As its involvement in Vietnam grew, the United States underwent more and more changes in self-perception.

Spurred by quantum leaps in technology, a series of revolutions — mostly peaceful — shook society and many of its attitudes. The sociological revolution was spawning a counterculture known as the hippie movement, a drive for equal opportunity, and new public ideas about morals and manners. The racial revolution was nearing its crest, and education had its own revolution.

Young people around the world reached out for more learning, got it, and demanded a say in how it should be delivered. About 50 percent of college-age Americans were enrolled in school, compared with only 4 percent at the turn of the century.

The communications revolution, spearheaded by the computer and TV's "windows on the world," gave many millions of people a new focus on their planet, quick exposure to fresh knowledge, and an understanding, sometimes coupled with envy, of how others lived. A new psychology of entitlement was spreading over the world: people not only expected more for themselves from society; they demanded it.

The transportation revolution was equally dynamic. Everyone wanted wheels, and most got them, transforming the look of landscapes in the United States and abroad. Motorways, autobahns, and autostrada fanned out from metropolitan areas and between cities to create new living modes in both suburbia and exurbia. The whir of tires on roadways and the throb of jet planes overhead replaced the lonely, cozy sounds of railroad whistles in the night.

Goodyear rode the crest of the changes, racing to new corporate records. In 1964, it became the first company in the rubber industry to exceed $2 billion in sales when they mushroomed to $2.0106 billion. This was 16.1 percent greater than in the previous record year of 1963, making Goodyear the 21st largest industrial corporation in the United States. Net income soared past the $100 million level to an industry record of $100.1 million.

Chairman DeYoung, obviously relishing his first year as CEO, said: "It took 53 years, from 1898 until 1951, for Goodyear to reach a $1 billion sales level and only 13 more to pass the $2 billion mark."

Attesting to the increasing importance of operations outside the United States, profits of foreign subsidiaries amounted to $35.9 million in 1964.

DeYoung reported that capital expenditures had averaged nearly $100 million per year for 1962, 1963, and 1964 and that the total for the 10-year period through 1964 exceeded $800 million.

The company was, he said, "in an excellent position to forge farther ahead in 1965."

## Community in the Desert

With 1964 profits of foreign subsidiaries totaling more than a third of the corporate total, no one doubted a good part of the "forging" would be outside the United States. Less than three weeks after the report of record sales and profits, DeYoung disclosed that capital expenditures of $150 million would be authorized for 1965 and "slightly more than half of the total covers domestic projects." The remainder would be appropriated for international facilities.

As if to underscore the determination of DeYoung and his management to increase tire-market share in the United States, the company announced the largest single construction project in its history in 1967, a $73 million tire plant on a 593-acre site at Union City, Tennessee.

When the Union City announcement was made, construction of a different nature was progressing gradually 2,000 miles to the west where Goodyear was building a new community in the Arizona desert. With an Arizona population explosion in general and growing appeal for full-time residents, vacationers, and business enterprises in particular, the city of Phoenix was spreading in all directions and had reduced from 16 miles to 6 the distance between its outskirts and Goodyear Farms, the agricultural subsidiary in Litchfield Park.

*The Wigwam in 1937 during its early growth from a small company hotel to a nationally acclaimed vacation resort.*

This 22,000-acre property, acquired by Goodyear in 1916 for cotton growing, now included the farms, the community of Litchfield Park, and the nationally recognized vacation resort, the Wigwam. It was dotted by several hundred residential lots sold during the '40s and in 1950, mostly to company employees and professional people who supplied services to the community.

In 1962, the company formed the Litchfield Park Land and Development Co. as a subsidiary of Goodyear Farms to provide a complex of 12 villages. These settlements would form six communities around a central area of about three square miles. In 1964, a $5 million program was initiated for residential, commercial, and industrial development of about 2,100 acres of the 12,000 remaining acres on Goodyear Farms. The Wigwam's 18-hole golf course was remodeled and expanded in 1965, and construction began on two additional courses designed by Robert Trent Jones, the noted golf course architect. One new community was developed around the golf courses, another around a grouping of man-made lakes.

Goodyear directed the long-range development plan toward a community that within 25 years should provide attractive living for a population of up to 100,000. The arid desert land purchased for the company so many years ago by P. W. Litchfield had proven its value many times, even though in 1920 it had been a heavy financial burden when cotton prices plummeted. The property was now assuming a much different character and appearance than had been envisioned by Litchfield a half century earlier.

The 50-year change in Goodyear's desert landscape was no greater than in the look of the corporation itself. In 1916, Goodyear had become the largest tire company in the world, but sales volume was less than $100 million. Although it operated a tire manufacturing subsidiary in Canada, had sales branches in Australia, Argentina, England, and South Africa, and had just acquired its first plantation in Sumatra, 90 percent of its interest and effort was focused on the U.S. market.

Except for its Bowmanville factory, nearly all manufacturing was concentrated in Akron. The Goodyear course was directed by a handful of men there known personally to a majority of employees, and it was logical and easy for Goodyearites to feel and share the family spirit expressed in the title of its publication, *Wingfoot Clan*.

Fifty years later, Goodyear was a multinational behemoth, ranking among the 21 largest U.S. industrial corporations, and had operational interests in almost every country outside the Iron and Bamboo curtains. Its 1966 annual report listed in excess of 240 products made at 97 facilities. The worldwide payroll supported more than 108,000 employees, but thanks to the computer revolution, many shared their duties with spiritless machines.

## Goodyear Spirit Recognized

Could the Goodyear spirit, the spirit of the clan, be kept vital in such a mammoth, widely dispersed, and increasingly mechanized organization? That concerned E. J. Thomas, now in his second year after retirement as CEO and board chairman.

*Mr. and Mrs. E. J. Thomas (left) with Florence McGowan, "the voice of Goodyear" and winner of the first Mildred V. and Edwin J. Thomas Goodyear Spirit Award, 1966.*

On May 26, 1966, the 50th anniversary of his joining the company, Thomas and his wife announced the establishment of an annual Mildred V. and Edwin J. Thomas Goodyear Spirit Award. This award was to be given each year to the Goodyearite who best demonstrated significant contributions to the "creation, maintenance, and growth of what has come to be known as the 'Goodyear Spirit.'" It consisted of a bronze medallion and a $1,000 cash grant, which was increased to $1,500 in 1978.

Any employee, active or retired, would be eligible for the award, and nominations could be made by any employee or retiree. In later refinements, Goodyear Spirit Awards were given to divisional and subsidiary winners who had qualified for final judging by a top-management panel.

In describing the award, Thomas said:

> Goodyear spirit is something Goodyear people have always had to a marked degree, and it has been outstanding in our or any other industry. While it is difficult to define, we know it includes the demonstration of constant and faithful loyalty, devotion, allegiance, and fidelity to the company and its people . . . It calls for the exemplification and regard for human qualities. It means to imbue with spirit, ardor, fire — for our company beyond the normal call of duty.

The first Goodyear Spirit Award went to Florence H. McGowan in 1966, a telephone operator and "the voice of Goodyear" in the New York district sales office for nearly 40 years. Chairman DeYoung presented the award at the December 22 employee Christmas party in the Goodyear Theater. E. J. and his wife were in the audience.

## A Series of Good Years

If the Goodyear spirit seemed more evident than usual in 1966, there was good reason. The economies of most industrialized nations were healthy, and automotive industries on all fronts were making good progress. Goodyear was progressing with them — perhaps was even a bit ahead of them.

In February, the company reported sales and net income records of $2,226.2 million and $109.2 million for 1965. DeYoung pointed out that the company had supplied a record volume of tires to the auto industry and its replacement market during the year, and "demand for tires and the thousands of other products manufactured by Goodyear resulted in capacity or near-capacity operations by our production facilities throughout the world."

*Shepherd with flock outside Goodyear plant in Izmit, Turkey. Ownership in the Turkish subsidiary was offered to the public, a departure from 100% company ownership of foreign plants.*

*Above, his Royal Highness Grand Duke Jean of Luxembourg (left) at inauguration of Luxembourg fabric mill, 1969. Top right, President Mobuto of Zaire cuts the ribbon opening Goodyear's new plant in Kinshasa, as Goodyear Chairman DeYoung (center left) looks on, 1972. Right, President Suharto of Indonesia (center left) and DeYoung (far right) at dedication ceremonies for expanded tire manufacturing facilities in Bogor, Indonesia, 1971.*

He added that the company had carried out the largest capital expenditure program in its history, with total authorizations at $195 million. When asked why Goodyearites should consider 1965 the best year ever, he answered, "Because they felt it in their pocketbooks." Wages and salaries that year had risen 9.3 percent, employee benefits nearly 20 percent.

Although 1965 was indeed "a very good year," 1966 was even better. Sales and earnings again reached new highs, gaining 11.2 percent and 8.5 percent over 1965 records, and the increases were universal. Earnings of foreign subsidiaries were more than $36 million, or almost a third of the corporate total of $118.4 million. In the United States, 2 new plants were completed, construction of 2 others started, and 100 retail outlets opened. Outside the United States, 7 new plants were completed and 17 expanded or modernized.

The report to shareholders for that year said, "Demand for Goodyear products throughout the world was never higher," but "growth in the immediate future may be at a somewhat slower rate."

These predictions were borne out in 1967 with new sales and earnings records, despite two-week strikes at 11 U.S. factories. This was the sixth consecutive year of record sales, the fifth consecutive year of record earnings. But the increases for the year were not as healthy as those preceding them. Sales of $2,637.7 million were up 6 percent, earnings of $127 million up 7.2 percent. Shareholders were told, however, that "in the past five years sales volume increased over $1 billion, and profits during the same period increased by $56 million."

The onrush continued to the end of the 1960s. Sales gained 10.9 percent in 1968 and net income 16.7 percent. Then in 1969, Goodyear became the first company in the rubber industry to top $3 billion in sales as an increase of 9.9 percent brought the total to $3.2153 billion. Net income also reached a record high, rising 6.7 percent to $158.2 million.

Record foreign earnings of $48.5 million "reflected the results of substantial expenditures in recent years for expansion and modernization, as well as market growth in essentially all areas of the free world," the annual report stated. Capital spending during 1969 for expansion, modernization, and replacement throughout the world was at a high of $293.3 million. The lyrics of a popular song that described each passing year as "a very good year" could well be applied to the company's experiences in the '60s. It was a series of very good years.

*Goodyearites enjoy a sunny day at Wingfoot Lake Park.*

## Wingfoot Lake Park

Employment at Goodyear had grown to about 140,000 by the late 1960s, and about 10 percent of the total was in the Akron area. Despite the size and scattered domicile of the Akron group, it retained much of the old clan spirit. It was especially evident in the wide-ranging recreation program that brought thousands of employees together in shared interests or athletic competition, but few recreational activities were family oriented. With an eye on the large Goodyear property at Wingfoot Lake, 12 miles southeast of corporate headquarters, Chairman DeYoung decided to do something about that.

Chuck Bloedorn, director of recreation from 1957 to 1977, recalled: "Russ took me out to Wingfoot Lake in early 1968, and we made an inspection of the picnic-camping area across the lake from the Aerospace hangar. He didn't like what he saw. Shortly thereafter the Wingfoot Lake Park redevelopment program began."

This property had been Goodyear's for more than 50 years. In late 1916, the company purchased 720 acres in the Fitch's Lake section of Portage County for its balloon-building facilities. The lake also would contribute to the water supply for factory operations in the headquarters complex.

In 1917, when Goodyear was contracted by the Navy to build nine blimps, the company constructed a hangar on the south shore. Support facilities, such as a hydrogen plant, a cantonment, and workshops, also were put up, and the Wingfoot Lake site was used by the Navy as a training base for balloon and blimp pilots until the end of World War I.

In the postwar years and for more than three decades, the Goodyear site on the lake's northern shore, across from the blimp facility, was developed haphazardly into a picnic grounds and camping park for employees. The wooded area was dotted with open fields and offered a canteen, picnic tables, huts, some outbuildings, and children's play areas. "It certainly wasn't fancy," said Bloedorn.

DeYoung's redevelopment project was completed in less than three years and gave Goodyearites and their families a park of 75 wooded, rolling acres and a 522-acre lake for boating and fishing. Its facilities, continuously improved and expanded in succeeding years, included shelters with electric outlets and fireplaces, 300 picnic tables, grills, softball diamonds, and courts for boccie, basketball, volleyball, tennis, and badminton.

Shuffleboards, horseshoe pits, a hole-in-one (pitch-and-putt) golf course, rowboats and pedal boats, a putting golf course, canteen, rest rooms, four well-equipped playgrounds for children, and areas for the Goodyear Boy Scout troops were also provided. The lake, stocked annually by the Goodyear Hunting and Fishing Club and the Engineering Club, had a good population of catfish, perch, walleyes, bass, pike, and bluegills.

Park attendance in 1969, the first year for which attendance was recorded, was 24,000. It exceeded 90,000 by 1974 and 100,000 by the 1980s. More than 5,000 family reunions were held there from 1970 to 1982.

*The Eagle GT radial, designed for 1980 Corvettes and unveiled on the Indianapolis 500 Pontiac Trans-Am pace cars that year, is lowered onto a test wheel for a noise evaluation in the anechoic chamber at Goodyear's Tire Dynamic Test Laboratory in Akron.*

# Chapter XIII

# THE RADIAL AGE

The world's tire industry steamed ahead at full speed as the 1960s ended. Motor transportation expanded everywhere as economies grew, international trade flourished, and mankind insisted on greater and more efficient movement of people and goods. Auto registrations throughout the world had nearly doubled from 1950 to 1960, and truck registrations had increased 60 percent.

Rubber-tired motor vehicles operated in environments as diverse as the sands of the Sahara, the forests of Canada, the autobahns of Germany, the mountains of Peru, the swamps of Louisiana, and the crowded streets of New York City, London, and Tokyo. Technology labored to keep up with the demand for tires to provide optimum performance on each and all. It was most successful in Europe where the innovative radial tire was changing the marketing mix and beginning to shake the very foundations of the tire industry.

The radial was introduced in 1948 by Compagnie Générale des Etablissements Michelin of France. It was a radical departure from the conventional bias-ply tire, in which the reinforcing cords cross the tire's body at an angle, extending diagonally across the tire from bead to bead. In a radial-ply tire, the cords run straight across from bead to bead substantially perpendicular to the direction of travel, and an additional layered belt — of fabric, steel, or glass — is placed between plies and tread. From the start, Michelin's belts were steel.

Headquartered at Clermont-Ferrand, Michelin was France's leading tire maker and the seventh largest in the world. It ranked with Dunlop, Continental, and Pirelli at the top of the European-United Kingdom tire industry, and because of the radial tire would soon be the number one tire manufacturer in the European region.

This company had been a factor in the U.S. market, operating a factory at Milltown, New Jersey, in the 1920s. It was shut down during the depression of the 1930s, and Michelin's presence in the states became minimal, although its imported tires retained an excellent reputation for quality. Michelin probably is best remembered by older American motorists for its advertisements of the 1920s and '30s that featured Mr. Bibendum, the roly-poly man made of tires. Mr. Bib soon would become well-known to a new generation of American drivers.

Privately controlled and highly secretive, Michelin stole a march on the European tire industry by quietly developing the radial, then thrusting it on the European market in 1948 at premium prices as the Michelin X. This revolutionary tire was strongly supported by shrewd promotion, aggressive marketing, and a well-cultivated mystique of quality.

Michelin had always carefully guarded its know-how and production processes, even segmenting and separating various manufacturing operations so that workers in one division or section knew little or nothing about what other divisions or sections were doing. Whether the aura of secrecy was cultivated for promotional purposes or not, the mystique it created undoubtedly was a marketing plus, becoming almost legendary as the Michelin X became the most highly regarded tire in Europe.

One of the legends of secretiveness at Clermont-Ferrand concerned General Charles de Gaulle, France's heroic and sometimes haughty president. According to the story, while in the vicinity of Michelin's headquarters, de Gaulle expressed a wish to visit its adjacent tire factory. He was politely, but firmly, turned down. Perhaps the anecdote was apocryphal, but it and others of the same nature added to the Michelin radial mystique.

## Polyester Cord Tires

As the radial tire swept through Europe in the late 1950s and the 1960s, adding to Michelin's sales volume and reputation, two technological developments did the same for Goodyear in the United States, positioning it for the radial tire wave of the '70s: the polyester tire cord (early 1962) and the bias-belted tire (1967).

Goodyear had begun intensive evaluation of polyester as a tire cord in the late 1940s, along with several other U.S. tire companies. It was part of an ongoing search for cords to satisfy four major areas of need: a smooth, quiet ride to complement quieter cars rolling out of U.S. factories; a high-speed, cool-running tire for use on long stretches of expressways and turnpikes; resistance to cuts and bruises; and improved economy. In 1958, the company initiated an all-out program to develop new tire fibers.

Fabric reinforces tire rubber much the same as steel reinforces concrete. Pound for pound, though, most fabrics are stronger than steel and have the highest fatigue resistances among reinforcing materials. Cotton, the first fabric used in pneumatic tires, continued as the only tire fiber from its first application in the mid-1890s until 1938 when rayon, a synthetic fiber, was developed for use in tires. Nylon, another synthetic, joined the tire-cord family in 1947. Although not the first to use cotton — which was built into pneumatics five years before the company was founded — Goodyear had pioneered the use of long staple cotton, rayon, and nylon and in 1962 took the lead in promoting polyester tire cord.

After exhaustive testing of polyester, Goodyear's research and development engineers, headed by Walter J. Lee, director of tire research and development, were convinced it was closest of any known material to being an all-purpose tire fiber. Classed in the same strength category as nylon, polyester proved itself more efficient and economical as a tire material than either rayon or nylon. Its dimensional stability called for the use of less fabric in a tire, and its specific gravity allowed the use of less rubber.

Goodyear introduced the first commercial polyester tires in April 1962. Still more costly than nylon, Goodyear's polyester cord — known as Vytacord — was first used commercially in the premium Double Eagle tire. As more Vytacord became available through expansions at the company's Point Pleasant, West Virginia, plant, the new cord was extended to use in more tire models. By the end of '62, it was in 8 of the company's 14 auto tire lines. In 1967, the Polyester Technical Center was established in Akron as a new wing of the Research Laboratory, devoted exclusively to polyester R & D.

The Double Eagle with Lifeguard Safety Spare, designed to provide the utmost in road safety, went on the market in 1963. The 1962 Double Eagle tire had a steel cord Safety-Shield and required a second valve in the sidewall to permit inflation of the outer air chamber. The new tire was inflatable through a single valve in the rim through which air was simultaneously distributed to both chambers.

The Lifeguard Safety Spare had two parts, a two-ply cord body with its own beads and tread and a separate rubber air container. A built-in safety signal informed the driver of puncture or other inflation problems by causing a slight thumping. The driver then knew that driving without danger of further tire failure was possible for nearly 100 miles, but repairs were necessary.

During the winter of 1963-64, studded tires were introduced on a test-market basis. The company's regular Suburbanite snow tires were modified to contain about 100 tungsten carbide studs protruding about 1/16 inch from the tread

*Radial tire production at Goodyear's Gadsden, Ala., plant.*

surface. The studded tires gained quick acceptance by motorists, but aroused resistance in several states because of the possibility of damage to roads and highways. Within a few years, most states had outlawed studded tires.

Drawing on the development know-how gained in motor racing, Goodyear introduced two new high-performance car tires in 1966 that incorporated some of the features of racing tires, the Speedway Blue Streak and the Speedway Wide Tread. Engineered specifically for street use, they had a racy look and provided improved handling for sports cars.

U.S. tire sales in 1968 were supported by the largest package of TV advertising in the industry's history when Goodyear cosponsored both the winter and summer Olympic Games telecasts over the ABC-TV network. An estimated 140 million viewers watched telecasts from Grenoble, France, of the Winter Games in February and from Mexico City of the Summer Games in October. Goodyear offered a total of 66 minutes of commercials to these viewers, the largest concentration of TV commercials by a tire manufacturer up to that time.

As polyester cord tires were gaining acceptance by U.S. motorists and Detroit automakers alike, the motoring public was developing new tire-performance expectations: improved handling; greater stability; faster response; no thumping; less ride roughness and harshness; less tire squeal, whine, and rumble; and — to become paramount in the mid-'70s — lower rolling resistance and greater fuel economy.

Some of these expectations, especially handling, stability, rolling resistance, and fuel economy, were probably influenced by imported European cars shod with radial tires, to which Goodyear responded with the bias-belted tire.

Michelin had the radial tire field much to itself throughout the 1950s. By 1960, though, major European and British tire makers also were manufacturing radials, and in Luxembourg, a tiny segment of the world's largest tire and rubber company had already started its own attack on the surging radial market.

## Tire of the Future

Goodyear's first experimental radial automobile tires were made in 1958 at the European section of its International Tire Development Department, later called the Goodyear International Tire Technical Center, and Goodyear first made radial truck tires there in 1960. Goodyear's early radials were produced in primitive conditions on crude machinery, reminisced E. W. Williams, one of the seven technicians who staffed the Luxembourg center at its founding in January 1957.

> A radial passenger tire-building machine was purchased from a competitor and installed in our first Experimental Shop, located on the second floor. The great day came when Dr. R. P. Dinsmore, vice-president of research and development, was to visit, and we would make a radial tire-building demonstration.

> Our best — and only — tire builder, Aloyse Bertholet, proceeded to build a perfect tire on the machine. He then discovered it would not come off the drum. Harry Schroeder, the center's director, tried to persuade Dr. Dinsmore the demonstration was over, but Dinsmore insisted on staying to see if Aloyse could get the tire off the drum. Aloyse finally got it off, but it was not worth curing.
>
> From this "success," we went on to developing radial truck tires.

Four years later, Goodyear was producing steel-belted radial passenger and truck tires in small volumes at Luxembourg, and the world's largest tire manufacturer was officially in the Radial Age.

The radial wave would not sweep full force across the Atlantic for nearly 10 years. In the late 1950s and early '60s, the United States was not ready for this new tire. American drivers were accustomed to the soft, cushiony ride provided by conventional tires, and that is what they liked best. Owners of sports cars willing to pay the higher prices switched to radials because of increased tread life and improved traction, handling, and cornering. Most U.S. motorists, though, were not as concerned with these factors as the Europeans, who drove on narrower roads and fewer expressways and did not face the same restrictive speed limits.

Additionally, radial and conventional tires could not be used together on a car. So, replacement radials had to be bought four at a time — which turned off cost-conscious motorists.

Detroit was not eager to encourage radials and their higher costs; nor was it happy about the prospect of changing car suspensions to accommodate the radial, because that too would boost manufacturing costs. The U.S. tire industry, tied to Detroit's apron strings, shared these apprehensions. A conversion to radials would mean heavy investment in new tire-making equipment, an increase in tire-building expenses, and greater demand for accuracy and consistency in the one-ply radial than in the two-ply conventional tires.

Production managers in particular were urging caution by manufacturers before jumping from the conventional bias tire to the harder-to-make, more-costly tire of the future, as the radial was called in the 1950s and early 1960s.

The U.S. tire industry faced a dilemma. Should it play a waiting game to determine if the radial really was the tire of the future before developing competitive radial technology and making the costly conversion to radial equipment? Or should it abandon the conventional tire and go all out to gain dominance in the still small radial tire market, hoping Detroit would quickly adopt the new tire in response to public demand?

The question was one of timing: when to go radial. Goodyear's decision — which later proved Solomonic — was to develop the bias-belted tire, a hybrid between the conventional and radial tires, and support it with a massive marketing effort while speeding up its radial tire technology drive on both sides of the Atlantic and pushing hard in the swelling European market.

## A Transitional Tire

In November 1967, the company introduced a bias-belted tire that combined bias plies, belts, polyester, and fiberglass with a new wide tread. Goodyear claimed the new tire would save gasoline, grip the road better, and give up to 50 percent more wear than conventional tires. It was named the Custom Wide Polyglas and had a bias-ply Vytacord body with two fiberglass belts under the tread, which was about two inches wider than in the conventional tires.

Although major auto suspension changes were necessary to accommodate the radial tire, U.S.-made autos could be adapted to the bias-belted tire with only minor modifications. Goodyear's new tire was an answer to Detroit's prayer, and six months later the new concept was extended to sizes that would fit all standard American cars.

The Polyglas generated such heavy demand from automakers and motorists that the first million units were produced in five months and the second million in three. When Chairman DeYoung announced the new tires as standard or optional equipment on most 1969 model autos, he said no other tire had gained public favor so rapidly. Apparently it was just the kind of tire needed during the transition from conventional bias tires to radials, and it helped move Goodyear farther ahead of its domestic competition.

Goodyear also began national marketing of a radial tire in 1967, the Power Cushion Radial-Ply, after test marketing it in nine cities a year earlier. This tire provided 50 to 100 percent improvement in tread life and less rolling resistance, according to Charles Eaves, vice-president of sales. But U.S. motorists, paying only about 33¢ for a gallon of gasoline and unconcerned with fuel efficiency, displayed little interest.

The radial tide moving across the Atlantic was inexorable, though, and the U.S. automotive and tire industries had little chance of stopping it. This tide would combine with the cohesive actions of a group of oil-producing nations to reshape the U.S. tire industry during the 1970s.

## A New Breed of Manager

Few Goodyearites were more acutely aware of the merits of the radial tire and the need for Goodyear to go radial than Charles J. Pilliod, Jr., who in 1963 transferred from Brazil, where he had been commercial manager for three years and managing director for four, to Great Britain as sales director for one year and managing director for two.

In his British assignment, Pilliod had seen the radial's rise to full ascendancy in the European-United Kingdom market and had pushed hard for the development there of Goodyear radials. He continued that push in the United States after his 1966 return to Akron. In 1967, he succeeded Goodyear's Mr. International, Sullivan Kafer, who retired as a vice-president of the international subsidiary.

Sully Kafer and Chuck Pilliod personified the new breed of multinational executives whose global perspectives had been developed through long years of service abroad. Kafer, a Goodyearite since 1922, began his international career with the old Goodyear Export Co. in 1933 as service manager in Paris. From 1939 to 1942, he sold Goodyear products throughout the Mideast and North Africa.

He was manager of Goodyear-Colombia from 1943 to 1947 and the managing director in Indonesia until 1956, when he returned to Akron as manager of the western hemisphere division. He became managing director in Brazil in 1958, then regional manager for Europe, the Near East, and Africa for two years. In 1961, he was regional director for Europe and, in 1962, director of operations for Goodyear International, moving up to the vice-presidency in 1964. On his retirement, the Akron *Beacon-Journal* published a full-page feature on "Goodyear's Modern-Day Marco Polo."

*The Custom Wide Polyglas tire. Introduced in 1967, Goodyear's bias-belted tire helped ease the transition from the conventional bias to the new radial tire for U.S. motorists. No tire had ever won public acceptance so quickly — the first two million units were produced in only eight months.*

Pilliod began his Goodyear career in 1941 on the training squadron. After serving as an Air Force pilot in World War II, he returned to the company in 1945 on the sales staff in international operations.

From 1947 until 1966, Pilliod held numerous foreign assignments. He was managing director in Panama; field representative in Peru, Chile, and Bolivia; sales manager in Colombia; commercial manager and managing director in Brazil; sales director, then managing director, of Goodyear Great Britain, the company's largest non-U.S. subsidiary. He was brought back to Akron in 1966 by Richard V. Thomas, president of Goodyear International, as director of operations and Thomas's right-hand man and troubleshooter. Eight years later, he would be elected Goodyear's fifth board chairman.

In the year Pilliod returned to corporate headquarters, the man who would succeed him as Goodyear president in '74 took an important step forward in the management ranks. John H. Gerstenmaier was elected vice-president for production in August, succeeding Walter H. Rudder, a 38-year production veteran, who was elected an executive VP and board member. Gerstenmaier, who had become director of domestic manufacturing after three years as president of Motor Wheel, had extensive experience in industrial products development and was formerly manager of the Logan, Ohio, plant, which manufactured automotive trim materials. In 1952 he had been awarded a Sloan Fellowship to MIT.

## 'Daring and Imagination'

In a July speech to members of the 1967 training squadron, Chairman DeYoung noted that Goodyear had an average investment of $20,102 in each employee and urged the recruits to help make the investment pay off through "daring and imagination," as he said.

> Today as you embark upon your career with Goodyear, there never has been a greater premium on daring and imagination . . . This is an outfit where daring and imagination are a way of life . . . We started in 1898 on $13,500 of borrowed capital. By 1916 we were the number one tire company in the world.
>
> By 1926, we were the number one rubber company. We have never been ousted from either position — nor do we intend to let that happen . . . It has taken a lot of effort by a lot of people to keep us on top. These individuals all have one thing in common. They are self-starters — which is another way of saying they have daring and imagination, plus aggressiveness.

He explained that Goodyear's sales had increased $1 billion in the previous five years and that 17,000 people had been added to the payroll to handle the new business. Citing the company's increasing diversity as a source of opportunities for employees, he said:

> Goodyear is more than a tire company, even though it is the biggest tire company and makes tires for every type of vehicle from garden tractors to jet airliners.
>
> It is more than a rubber company — the world's largest rubber company — even though it produces everything from tiny

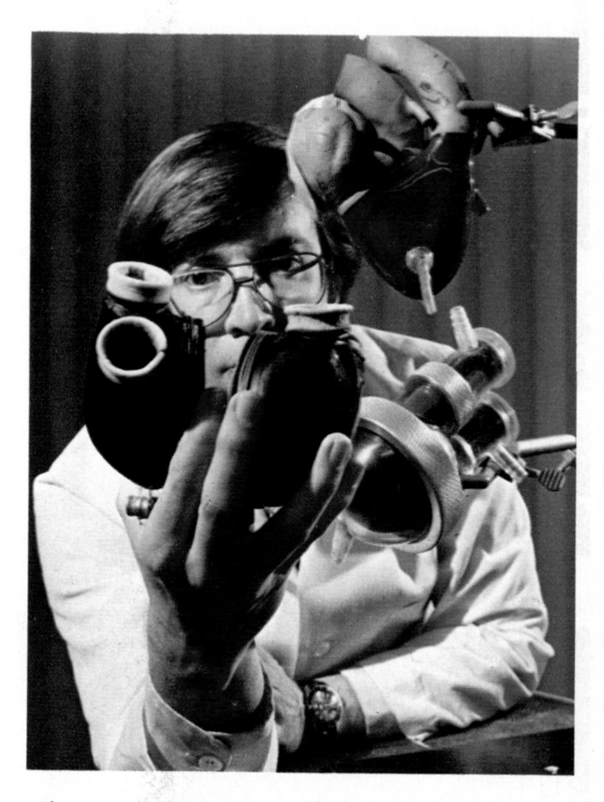

*Above, 15 years of designing artificial hearts led Goodyear researchers from the bulky, handcrafted model of the '60s (background) to this 1976 version made of compression-molded synthetic rubber and small enough to be implanted in a human. Opposite, "a preview of a future without traffic jams," the Peoplemover at Disneyland, introduced to the public in 1967.*

> rubber V-belts used in home appliances to huge rubber suction hoses for dredging rivers and streams . . . The rewards we offer are great — not alone in terms of financial success, but in a capacity for personal accomplishment and service.

In the same month, a product of Goodyear "daring and imagination" was unveiled at Disneyland in California: a new automated transportation system, the Goodyear People-Mover. It consisted of 62 four-car trains propelled by electric-powered rubber wheels and could carry 4,885 Disneyland visitors an hour to and through many of the Tomorrow-land pavilions. Synchronized revolving transform platforms enabled passengers to get on and off while the cars continued to move.

Not all the new products and product improvements brought forth by Goodyear daring and imagination in the 1960s were as forward looking as the PeopleMover, which company publicists saw as "a preview of a future without traffic jams."

But the list was long and diversified, ranging from an artificial heart to flotation bags for the earth-orbit missions to polyester tires. The spirit of innovation that had produced the straight-side detachable tire in 1903 — "the tire that made Goodyear" — seemed alive and well in Goodyear's wide world.

*Apollo 14's MET (Modularized Equipment Transporter) shod with Goodyear's XLT tires — the first tires on the moon. Photographed from inside the lunar module at its landing site on the moon in February 1971. (Courtesy of NASA)*

# Chapter XIV

# FLY ME TO THE MOON

As the Auto Age rolled forward, consuming more Goodyear tires than any other brand, the Space Age revolutionized the work and products of the company's aerospace subsidiary. Its name was changed in 1963 from Goodyear Aircraft Corp. to Goodyear Aerospace Corp. to reflect more accurately its role in production of aerospace and military items for projects that included missiles, guidance systems, radar systems, aerospace ground support equipment, flight simulators and trainers, and space satellites.

GAC's engineering and production versatility in the late '60s was illustrated by some of the products it supplied for the space program and the military services. The world's largest space brake, the Ballute, is a good example.

Built for the National Aeronautics and Space Administration (NASA), the Ballute was three stories high and shaped like a top. It was designed to slow down space vehicles as they reentered the earth's atmosphere.

Another innovation was the largest antenna ever built for space. When deployed, it blossomed from its container to a diameter of 30 feet.

More items included:

A flight simulator for training Navy pilots to land on decks of aircraft carriers;

For the Marine Corps, nearly 1,000 collapsible Pillow Tanks to store liquid fuel, most with a capacity of 10,000 gallons;

For the Air Force, an ultrahigh speed device, the Associative List Searcher, which would search a list of 32,000 items in less than two one-millionths of a second;

For the Boeing 747B jetliner scheduled to fly in late 1969, 9,000 double-pane acrylic plastic cabin windows.

GAC equipment and technology had vital roles in man's first and second landings on the moon via *Apollo 11* and *Apollo 12*. Their contributions included the brakes for the giant transporter that moved *Apollo* missiles to the launch pads, the purge and conditioning system that carried outside air and heated nitrogen into the missile engine compartment and purged excessive concentrations of hydrogen and oxygen, and window frames for the command module.

Also in that list were the Bondolite panel on which landing module instruments were mounted, micromodules that handled signals for controlling the pitch of the *Saturn* missile, and flotation bags that the astronauts inflated from inside the command module to upright the space capsule after upside-down landings in the sea.

From its incorporation as Goodyear Aircraft Corp. in 1939 until 1956, GAC was headed by Paul Litchfield as president. Three presidents held the reins during its gradual transformation into a true aerospace company. Thomas A. Knowles, its wartime sales vice-president and postwar vice-president and general manager, held the post from 1956 to 1965.

Loren A. Murphy, who came from the parent company in 1925 and became vice-president of manufacturing, planning, and personnel in 1960, was president from 1965 to 1968. He was succeeded by Morris B. Jobe, who had joined the parent company in 1938, transferred to GAC in 1941, was its sales manager in 1960, then general manager of the Arizona division.

## Toward the Turbulent '70s

Goodyear stepped from the 1960s into the 1970s with an upbeat spirit and high hopes for the future. The record sales, profits, and capital outlays of 1969 had augured well for continued growth in the coming decade.

New production capabilities in international markets, particularly Europe and Asia, positioned the company to take advantage of mounting demands for more and better motor transportation. Auto production and registrations throughout the world outside the U.S. had doubled in the '60s, and trucks had kept pace. In the mature U.S. auto market, production had remained static during that period, but car registrations were up nearly 40 percent, and replacement tire sales had increased more than 40 percent.

America's mood was ambivalent, a mixture of hope and concern. National morale had soared in the summer of 1969 following the success of *Apollo 11*'s moon mission and Neil Armstrong's "one small step for man, and one giant leap for mankind." But the Vietnam War brought on an unsettling and divisive factor.

Dissidents disrupted or paralyzed college curricula. Environmentalists raised questions over pollution of the atmosphere and stimulated regulations that inhibited part of the industrial process. The radical underground was active and vocal.

The U.S. youth population, which had reached 40 million in the 14-to-24 age group, was making rock music, drugs, and the "liberated" life-style symbols of rebellion against the establishment, particularly against U.S. involvement in Indochina. The crime rate was high throughout the country; Martin Luther King and Robert Kennedy had been assassinated in 1968; and 71 commercial airliners were hijacked in 1969. These were troubled times for the United States.

Despite a sluggish and uncertain U.S. economy, the tire and rubber industry forged ahead. It was closing a decade of record growth, and its leader, Goodyear, had doubled sales in this 10 years. Although domestic tire companies

foresaw upheavals caused by the radial tire soon to hit home, most leaders looked ahead with confidence. The one thing they did not see, though, was the oil crisis of the mid-'70s that would shake their industry to its roots.

Goodyear's last major capital investment of the '60s was in a polyester tire yarn plant at Scottsboro, Alabama, a $12 million facility with 100,000 square feet of floor space on a 275-acre site. Yarn spun there would be shipped to the company's textile mills at Decatur, Alabama, and its three sites in Georgia: Cartersville, Cedartown, and Rockmart.

In late 1969, the company caught the eye of the construction industry by unveiling a giant plastic greenhouse at Wooster, Ohio, as part of a program to explore possibilities of air-supported enclosures for a variety of purposes. The greenhouse was 100 feet wide, 428 long, and 50 high and covered one acre. It was assembled in a small fraction of the time needed to build a normal greenhouse, and at one eighth the cost. The construction concept was to use tough, flexible film held in shape by steel cables and pressurized slightly by a blower to form the enclosure.

*Opposite, Apollo 11 astronauts prepare to enter life raft after successful splashdown, 1969. Goodyear-made flotation bags at top were used to upright the space capsule. Above, duplicate of "moon tires" that equipped the MET, the rickshawlike cart that carried equipment for Apollo 14 astronauts as they explored the lunar terrain in 1971. Model of MET at right.*

In the same year, the University of Akron announced that its new $11.7 million cultural center would be named the Edwin J. Thomas Performing Arts Center in honor of Goodyear's former board chairman. Following his retirement in 1964, Thomas immersed himself in community activities, particularly the rebuilding of downtown Akron and the growth of Akron University. The center was opened in October 1973 and received high acclaim as one of the finest performing arts structures in the nation.

## A Tire on the Moon

Goodyear's eight-year string of record-breaking sales was interrupted in 1970 when strikes in the rubber, automotive, and trucking industries forced U.S. tire factories to operate well below capacity. Goodyear sales were about $20 million below the record of 1959, and net income was down by $24 million.

Chairman DeYoung, suggesting that the overall economy was showing signs of "emerging from its doldrums," predicted that 1971 would be a better year. For Goodyear it was, as record sales and profits once more were achieved. Although profits shifted erratically in the '70s, the year-to-year sales increases would continue through 1982.

On January 1, 1971, Edwin J. Thomas, at 71, resigned as a member of the board of directors and chairman of its finance and executive committee. He had retired as chairman and chief executive on April 6, 1964, a year sales were 30 times greater than when he joined Goodyear as a stenographer in 1916.

Undoubtedly the idea of a Goodyear tire on the moon was beyond the wildest imaginings of young Eddie Thomas when he joined the company. Nevertheless, on February 5, 1971, six weeks after cutting his official ties with Goodyear, he and millions around the world watched telecasts from the moon as astronauts Alan Shepard and Ed Mitchell of the *Apollo 14* mission pulled a two-wheeled rickshawlike cart along the lunar surface. It was equipped with two Goodyear tires.

This rickshaw, officially named NASA's Modular Equipment Transporter, had been boosted 250,000 miles into space and, while attached to the outside of the landing module, delivered from the moon-orbiting command module to the moon's surface. Pulled by the two astronauts prowling the unexplored terrain, the wheeled cart carried cameras and film, shovels, scoops, and core tubes — plus 35 numbered bags into which samples of lunar rock were placed.

The moon tires, only 16 inches high, were called XLT, for experimental lunar tires. They each had an inner tube and were fitted on an eight-inch diameter aluminum wheel. In all, tire, tube, and rim weighed only 4.1 pounds.

Hundreds of Goodyearites had worked on the moon-tire project, which went through 10 years of changing designs as new theories were developed about the moon's surface and environment.

The variety of skills involved is indicated by the different departments that took part: in Akron, program management, tire design research, tube development, compound development, tire reinforcing systems, research division, physical test lab, machine design shop, industrial tires, Goodyear Aerospace materials lab; in Gadsden, tube fabrications; and in New Bedford, valve fabrication.

*Western-style boots are fitted with soles specially compounded with cork by Goodyear to reduce weight and increase traction.*

The key Goodyear administrators of the project were John J. Hartz, vice-president of tire development; Walt Curtiss, chief engineer, tire design research; and Jack Davisson, manager of auto and special tire engineering and director of the lunar tire program.

After 33½ hours on the lunar surface where they set up scientific experiments with the help of the Goodyear-tired cart, Shepard and Mitchell blasted off to dock with the orbiting command vehicle controlled by Stuart Roosa. The tiny rickshaw and its first tires on the moon were left behind, perhaps for future reclamation by another crew of astronauts.

## Imports Threaten

A month before Goodyear tires landed on the moon, former bomber pilot Chuck Pilliod was elected a corporate vice-president and president of Goodyear International Corporation. He succeeded Richard V. Thomas, who had resigned to pursue personal interests on a full-time basis; Thomas had headed Goodyear International since 1961.

Nine months after gaining a corporate vice-presidency, Pilliod was elected an executive VP and a director of Goodyear. At the same time, John H. Gerstenmaier, vice-president of production, and Richard A. Jay, vice-president of general products, were elected executive VPs; each later would serve as vice-chairman of the board under Chairman Pilliod. Succeeding Gerstenmaier was Thomas F. Minter, who for four-and-a-half years had been director of manufacturing.

A cloud that had been building on the horizon of the U.S. tire industry finally took ominous shape in 1971. Imports had captured more than nine percent of the U.S. market; this contrasted with just three percent five years earlier. Expressing industry concern, Edwin H. Sonnecken, Goodyear's director of corporate business planning and research, pointed out that 1970 auto and truck tire imports had totaled 19.4 million units, more than four times the 4.3 million tires exported.

He explained that on the average every million tires imported meant 457 fewer tire industry jobs in the United States. Sonnecken also pointed out that in 1970, foreign cars accounted for 15 percent of the U.S. market, compared with 10 percent in 1959, and that in

California one of every three cars sold in 1970 was an import.

Further, greater gains abroad in output per worker and differences in labor costs were also major factors to be considered. Foreign labor costs per hour, including wages and benefits, ranged from $1.05 in Spain to $5 in Canada. In the United States, these costs exceeded an average of $6 per hour including wages and benefits.

Tariff differences and export subsidies by foreign governments accentuated the problem, Sonnecken noted. As the U.S. automotive industry and its suppliers and citizenry would painfully learn, imports would reach flood tide by the late '70s, making the 1971 influx seem just a trickle.

Goodyear had already felt the effect of imports in another area, shoe products. By 1971, imports had gained about 83 percent of the domestic footwear markets, thanks in good part to lower wages in Europe and shoe-styling leadership in Italy, France, and Spain.

The company's shoe-product sales barely held their own through the marketing of new products and strategically situated factories at Windsor, Vermont, and Madisonville, Kentucky. These plants were especially helpful in challenging imports because they had good access to the U.S. shoe industry, which was concentrated in New England, Missouri, and the South.

## Company with a Conscience

Sonnecken was one of three elected to newly created corporate vice-presidencies in January 1972. His title was VP, corporate business planning, and he was joined by Joseph E. Hutchinson as VP, product quality and safety, and John P. Kelley as VP, advertising.

Hutchinson, a Goodyear veteran of 26 years, had served as director of product quality and safety since 1969 after extensive experience in engineering, design, and development. Kelley had joined the company as assistant director of advertising in 1959 and became advertising director two years later. Sonnecken came on board in 1961. The three new titles reflected an increased emphasis on consumer relations and customer orientation.

In April 1968, a corporate vice-presidency for materials management had been established, reflecting the need for high-level coordination of the company's widely dispersed purchasing and distribution operations. V. Lawrence Petersen was the first to occupy the position, as vice-president for materials management. His responsibilities included purchasing, merchandise distribution and control, traffic, and distribution services.

Petersen joined Goodyear as a development engineering trainee in 1947, advanced through various posts in the purchasing department, and was appointed purchasing agent in 1959. He was made manager of general merchandise and materials control in 1963 and director of materials management in 1966.

As antiestablishment attitudes spread nationwide in the last half of the '60s — in step with and perhaps because of the growing protest against the Vietnam War — consumerism, the equal rights movement, and environmentalism kept pace. By the end of that decade, ecology was a paramount public issue, pushed by growing news media and public awareness of increasing pollution of the earth's atmosphere and oceans, lakes, and waterways.

According to a 1970 Gallup poll, about 70 percent of the population considered environmental problems the top issue facing the United States, and industry perforce gave new and serious attention to pollution control. In 1971, Goodyear reported it had spent more than $16 million to bring older factories up to the environmental standards of its new factories and would spend at least an equal amount to complete its pollution control program during the next five years.

The drive for tighter environmental controls exemplified industry's response to society's rising clamor for protection against government's often overzealous attempts to do the protecting. Excessive regulations had become a tangible problem for business and industry, and Goodyear, along with the entire rubber industry, was no exception.

Goodyear could no longer rely solely on the excellence of its products, aggressive sales and advertising programs, and its reputation as a good employer to support marketing efforts. It had to demonstrate vividly a social conscience, a commitment to the general welfare. Management faced new public relations problems and shouldered new responsibilities — different from those of the Seiberling and Litchfield days — and not just because of the company's size and its extension throughout the country and the world.

All industry, particularly in the United States, faced new and demanding public expectations. Soon the oil crisis, rampant inflation, and the invasion of the radial tire would beset the U.S. tire industry. The 1970s would be a telling time for Goodyear and its management.

Although U.S. tire and auto manufacturers kept a wary eye on the radial phenomenon in Europe, the new tires had not made a big dent in the U.S. market by the start of the '70s. Goodyear's bias-belted tires, however, had proved a tremendous success throughout the states. Announcing the new Power Belt Polyglas in early 1971, the 13th line of bias-belted tires, Goodyear reported it had produced more than 56 million bias-belteds in the slightly more than three years since introducing them.

By 1972, the bias-belted was dominant in the U.S. market. It accounted for 50 percent of all automobile tires sold in the United States, compared with 42 percent for the bias-ply and 8 percent for the oncoming radial. While making ready for the Radial Age, the company was making hay — healthy sales and profits — with the bias-belted tire, which, on hindsight, came to be known as the interim tire.

*Top, dock fenders protect both ship and dock. By 1971, Goodyear was the largest manufacturer of dock fenders in the rubber industry, leading to the claim that "more boats and trucks bump against Goodyear dock fenders than any other kind."*

*Above, helicopter fuel tank. Goodyear's crash-resistant, self-sealing tank, developed by the aviation products division and Aerospace, won the Army's outstanding Civilian Service Medal for the company in 1971.*

## Industrial Products: Mundane, Glamorous

Since Goodyear had become the world's largest tire company in 1916, the major part of its business had been tires, but industrial products were not neglected as U.S. industry grew rapidly in the two decades following World War II. The company's domestic sales of industrial rubber products topped the $1 billion mark in 1969, and, at the start of 1971, Robert E. Mercer, general manager of the industrial products division, predicted they would exceed $1.3 billion for the year.

They did. Hose sales and belting sales each made up about a third of the total. Also, as U.S. car production climbed to about nine million units in 1971, Goodyear's engineered products for automobiles jumped from about $101 million to nearly $140 million.

Just as the company's advertising accurately claimed that "more people the world over ride on Goodyear tires than on any other kind," its industrial products management claimed in 1971 that "more boats and trucks bump against Goodyear dock fenders than any other kind."

The first dock fenders installed in the United States were Goodyear bumpers put on a Conneaut, Ohio, dock in 1933. By 1971, this segment of the company's industrial product sales made Goodyear the largest manufacturer of dock fenders in the rubber industry. Goodyear fenders in the '70s were produced in Akron and at the Northern Ireland industrial products factory. They ranged all the way from 3-inch fenders to larger protectors for parking garage pillars, and up to giant 60-inch absorbers installed on docks that moored giant ocean tankers.

One of Goodyear's more glamorous nontire products gained national attention in 1971 when the Army's highest civilian award, the Outstanding Civilian Service Medal, was presented to Chairman DeYoung for the company's development of its crash-resistant helicopter fuel tank. The tank, which also self-sealed bullet punctures, was the product of five years of research and development by the aviation products division, the special products development department of the research division, and Goodyear Aerospace. This kind of fuel tank was installed in 887 helicopters produced by the Bell Helicopter Co.

In presenting the medal to DeYoung at the Pentagon, the U.S. Army Chief of Staff, Gen. William C. Westmoreland, said:

"In 20 major accidents suffered by aircraft equipped with the system, there have been no injuries or fatalities as a result of fire. Our crash investigators have determined that 50 percent of the aircraft in major crashes could have or would have burned if it had not been for the crash-resistant tanks." He called the system a major breakthrough in aviation safety.

To keep pace with the rapid growth of the packaging films industry, Goodyear's films and flooring division established a new $2 million polyvinyl chloride packaging film plant at Calhoun, Georgia, in mid-1971.

Later that year, two Goodyear Speedwalk passenger conveyor systems were installed at the Cleveland-Hopkins Airport, extending from parking areas to the main terminal. They provided no-wait transportation for up to 7,000 people in each direction. Speedwalk systems had been installed previously at the Los Angeles, San Francisco, and Montreal International airports and at London's Heathrow Airport. Additionally, the company's Speedramp systems, which moved on an incline — not, as the Speedwalk, on level areas — were operating at the Akron-Canton and San Francisco airports and at Standiford Field in Louisville.

The Speedwalk-Speedramp business was dropped in 1974 because Goodyear, providing only the engineering and belting, could not exercise sufficent control, particularly in the quality of purchased materials and labor for the installation.

In response to the increasing need for accelerated industrial products technology in foreign markets, GIC announced plans in 1971 to construct a $2.4 million industrial products technical center at Craigavon, Northern Ireland. Its mission was to supply research, development, and equipment engineering for all the Goodyear industrial products operations outside the United States.

## GITC: Multifaceted Mission

Since its establishment in the tire factory at Colmar-Berg in Luxembourg in 1957, Goodyear International's technical center for tires had gained a solid place in the company's worldwide research and development scheme and had grown from a work force of 7 technicians, one tire builder and an assistant, and 6 drivers to a total of 750 employees, of whom 280 were technicians.

In 1969, the global chain of tire manufacturing operations served by the Technical Center gained considerable length when its Luxembourg-based engineers and technicians moved into their new, ultramodern quarters at Colmar-Berg. The center was moved from its plant space to a completely new research and development building adjacent to the Colmar-Berg factory.

A proving ground, including test track, was also completed that year, just a mile away. The technical center was now one of the largest and most modern tire development centers in the world.

Its multifaceted mission was to develop tires for the company's international markets, establish quality standards for product lines, work closely with vehicle manufacturers on tire approvals, provide specifications and technical assistance to the tire manufacturing plants worldwide, and develop raw materials and sources of raw materials for tires. The center also worked at tightening its liaison with the research and development division in Akron and served as listening post — the company's technical eyes and ears — during Europe's radial tire revolution.

The personnel at GITC, representing 20 nationalities, were organized according to the various disciplines in which they worked: tire engineering, automotive engineering, fabric development, compounding, tire research, tire evaluation, tire engineering services, materials evaluation, and systems engineering.

Many shared the responsibilities of serving as hosts and guides to a steady stream of technicians, engineers, dealers, representatives of the auto industry, company executives, writers, and editors who visited the center each year from throughout Europe and countries as distant as Japan, Mexico, and Australia.

Since the start of tire manufacturing operations at Colmar-Berg in January 1951, the factory's capacity had grown from 540 tires a day to more than 10,000 a day by 1969. In 1972, the 184-acre complex at Colmar-Berg included:

- the factory;
- the technical center;
- test track and proving ground;

- a $13 million fabric mill built in 1967;
- a $5 million plant constructed in 1969 to produce tire molds;
- a $7.5 million factory to make steel wire for tire cord completed in 1970;
- a $2.7 million airplane tire development and test center constructed in 1972.

Employment at Colmar-Berg totaled 2,800 in 1972, making Goodyear the second largest employer in Luxembourg after the Arbed Steel Company. The annual Summerfest, a day-long festival for all Goodyearites and their families in Luxembourg, was attended by nearly 3,000. It was one of the largest parties — if not *the* largest — in the country every year.

In September 1970, Goodyear International announced plans to construct a tire factory in the Congo (later renamed Zaire). This $16 million installation would be the first tire-making facility and largest foreign investment in consumer products manufacturing in that country.

During the following year, plans were revealed for several big new undertakings.

- A $13 million tire factory in Morocco

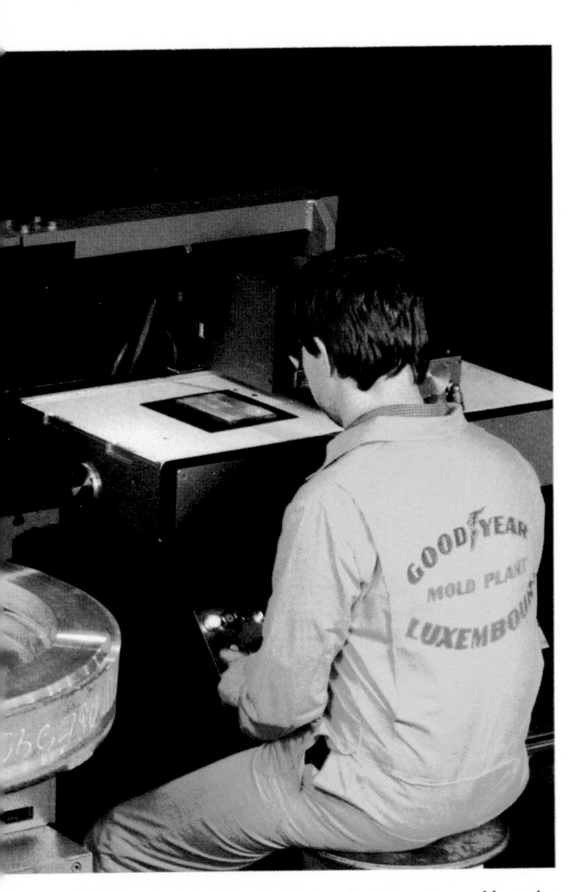


*Goodyear is engraved simultaneously in two tire molds at the company's mold plant at Colmar-Berg. The molds give tires their shape—and name—during vulcanization. The $5 million plant was added to the Luxembourg facility in 1969.*

- A $35 million tire plant in Brazil
- An Indonesian project in which a Goodyear International team would manage and operate two tire and tube factories owned by Indonesia's government
- A $13 million expansion of the Goodyear tire plant in Indonesia
- A $25 million expansion and modernization of the Luxembourg tire factory adjacent to the Technical Center

The Colmar-Berg expansion would increase the Goodyear investment in Luxembourg to $71 million.

The moves into Morocco and Zaire, the expansion in Brazil, and the agreement to manage the government-owned tire plants in Indonesia attested to Goodyear's increasing interest in the fast-growing tire markets of the Third World and its conviction that those markets held strong potential for even greater growth. Auto registrations in Africa and Asia, excluding Japan, had risen to 7.6 million in 1970, from 1.2 million in 1950, and in Latin America they had jumped to 6.5 million from 1 million.

As Goodyear's foreign investments of the next decade would demonstrate, management believed expansion of tire markets in the developing nations was only beginning.

## Indonesia: Fruitful Relations

The management project in Indonesia was a new kind of venture for the company. The five-year agreement to supervise and operate the Indonesian government's Intirub and Ban Palembang tire factories was the first agreement by Goodyear to manage a manufacturing facility for a foreign government. The two plants had been installed with East German assistance during the Sukarno regime. They were operating far below capacity because of insufficient technology and lack of trained personnel. After discussions with a number of companies from various countries, the Indonesian government chose Goodyear to provide know-how, train the factory employees, and bring both installations up-to-date in tire manufacturing. Undoubtedly the choice reflected Indonesia's confidence in Goodyear and an acknowledgment of the company's long and productive involvement there.

A Goodyear rubber plantation in Indonesia had been in operation since 1916, and the Goodyear tire factory at Bogor had started up in 1935 with 200 Indonesians producing 330 tires and 200 tubes a day. After its 1971 enlargement, the Bogor plant became the largest tire manufacturing facility in Southeast Asia, employing 1,600 Indonesians and daily producing 5,000 auto tires and 3,000 tubes, 12,000 bicycle tires and 2,300 tubes.

The Bogor expansion was dedicated in December 1971 by President Suharto of Indonesia and Russell DeYoung, who had spent 18 months in Indonesia in 1935-36 as a member of the team supervising construction of the original factory. In 1975, to symbolize the fruitful relationship between Indonesia and Goodyear and to express the Indonesian government's appreciation of DeYoung's long and active interest in Indonesia, its government presented the Bintang Jasa Nararya Award to the former Goodyear board chairman.

The intense competition among U.S. tire makers for larger market shares in industrialized countries was dramatized by the 1970-71 struggle between Goodyear and Goodrich to gain control of Vredestein, the largest Dutch tire and rubber manufacturer. The conflict began in April 1970 when Goodyear made a tender offer for Vredestein's shares. Goodrich, which had a large investment in the Dutch company including development facilities it did not want to pass into competitive hands, countered by increasing its investment. As a result, the two U.S. manufacturers controlled the preponderance of Vredestein's capital stock.

Subsequent litigation extended for about a year and indicated that neither company was able to exercise voting control of the Dutch company's affairs. So they reached an agreement under which Goodyear sold its shares to Goodrich at a $1.5 million gain. In the light of the coming, and largely unforeseen, overproduction of tires in Europe and a softening of the Continent's tire markets, this development was fortunate for Goodyear.

Key figures in the confrontation over Vredestein were Chuck Pilliod, president of Goodyear International, and John Ong, president of Goodrich's international division. Both later became board chairmen of their companies.

Twenty years after passing $1 billion in annual sales — the first tire and rubber company to do so — Goodyear also became the first in its industry to exceed $1 billion in compensation. Wages, salaries, and benefits for employees and retirees joined sales and earnings in reaching record levels in 1971. Total employee compensation was $1,014.4 million; sales were $3,601.5 million, a gain of 12.7 percent over 1970; and net earnings were $170.2 million, up 31.7 percent.

These increases represented a sharp rebound from 1970 when strikes at Goodyear's domestic plants and in the U.S. auto industry had a strong negative effect. The sales gain of slightly more than $400 million was greater than the combined gains of the other three members of the industry's Big Four. Firestone, the closest challenger, had an increase of $150 million; Uniroyal and Goodrich posted improvements of $120 million and $95 million respectively.

In the corporate surge to new worldwide records expressed in billions of dollars, Goodyear's management did not lose sight of the importance of the individual to overall progress. A two-page feature story in the Akron *Wingfoot Clan* of April 13, 1972, focused on Joseph Lord, an instrument repairman in Akron's Plant 5, as an example of "Goodyear employees who are still taking time to come up with better ways of doing things and earning sizable sums of money when they do." Lord was highlighted because he had just won $375 for an operating improvement suggestion for the plant's synthetic complex and had gained several previous suggestion awards since joining the company in 1943.

The news peg for the *Clan* story was in the first paragraph.

> Sixty years ago a Goodyear employee named Harry Thompson won the first cash award presented in the company's suggestion program . . . The $5 Thompson won undoubtedly was worth a lot more in buying power than now. But the most important fact of Thompson's payoff was that it signalled the beginning of a suggestion program that has grown into one of the most comprehensive and successful in the country.

In 1971, the year before Lord's award, 7,600 suggestions and ideas had been received by the Suggestion Department, and cash awards totaling $93,387 had been given for suggestions that resulted in company savings of $886,875.

## Pilliod Elected President

Thirty-one years after joining Goodyear, Charles J. Pilliod, Jr., was elected the company's 11th president by action of the board of directors, effective July 19, 1972. He succeeded Victor Holt, Jr., president since 1964, who would become 65 early in 1973.

Pilliod had been president of Goodyear International and an executive vice-president and a director of the parent company since 1971. In announcing the changes to a meeting of 300 managers in Goodyear Hall, Chairman DeYoung described Holt, a 43-year veteran, as "a man who belongs in the company of America's all-time sales geniuses."

In his early career, Holt had sales assignments in Akron, Kansas City, Oklahoma City, Miami, Harrisburg, and Philadelphia, then headed the tire sales departments in Akron. He was elected a corporate VP in 1956 with the responsibility for replacement tire sales and in 1958 became a director and executive VP.

Speaking of Holt's fame as a college athlete, DeYoung said:

> He was an all-American basketball player in the days when there was one all-American team that counted, and he won the Helms Foundation Award as the top basketball player in the country. Phog Allen, dean of American basketball, said of him, "Never outjumped, he outplayed every center he ever faced."
>
> Well, in our league, Vic Holt never has been outjumped or outplayed either.

## 'World is Our Market'

The advancement of Pilliod, who had spent 20 years in foreign posts and 6 in executive roles at Goodyear International's headquarters in Akron, in a way reflected the increasing multinational orientation and character of the company. Of his 10 predecessors as president, only E. J. Thomas and Russell DeYoung had experience overseas, and for both, the foreign assignments were for less than 2 years. Pilliod's global perspective and his frequently stated conviction that "the world is our market and the United States is just one of its segments" would have a profound influence on Goodyear's direction in the years immediately ahead.

Succeeding Pilliod as president of Goodyear International was Ib Thomsen, who had been a vice-president and director of the international subsidiary. A native of Denmark, Thomsen joined GIC's finance training program after receiving a master's degree in business administration from the Harvard Business School in 1952. He became a U.S. citizen the same year. He was a graduate of Niels Brock College in Copenhagen. After serving as a trainee with the Goodyear subsidiary in Sweden and as treasurer of Goodyear-India, Thomsen was assigned to Goodyear-Great Britain as treasurer in 1958.

In the next 10 years there, he was financial director and secretary, assistant to the managing director, deputy managing director, and managing director. He was elected chairman of the British subsidiary in 1968 and moved to Akron as GIC's vice-president in 1971.

Fifty-six years after its pioneering 1,540-mile round-trip journey between Akron and Boston to show the feasibility of pneumatic truck tires, Goodyear's Wingfoot Express rolled again. On July 19, 1973, a restored 1916 Packard truck with the same markings as the original pulled out from the company's headquarters on the first leg of the Akron-Boston run to the send-off of a horn-tooting, blank-pistol shooting crowd and the release of hundreds of helium-filled balloons.

It would follow much of the route covered by a similar vehicle in 1917 when most of the nation's one million trucks were limited to interurban runs because of solid rubber tires that minimized both speed and gasoline mileage and provided a jolting, teeth-shattering ride.

The first express had proved the use of pneumatics on trucks by making the return trip from Boston in 5 days, after 18 days, 23 hours, and 30 minutes on the outward run.

## Anniversary Express

The 1973 version had nothing to prove. It was part of the company's year-long 75th anniversary celebration. The recreated express made the 750-mile trip to Boston without a tire change, unlike the original, which had experienced blowouts about every 75 miles.

Goodyear's anniversary express made scores of stops en route, mainly for publicity purposes, and visited such cities as Pittsburgh and New York. In Philadelphia, the 1973 version made some history of its own by being the first vehicle ever driven on the mall in front of hallowed Independence Hall.

On the same day the replica express truck and its cavalcade rolled into New York accompanied by news photographers and television cameras, Goodyear reported record first-quarter and first-half sales and earnings. At year's end, sales were at an all-time high and earnings the second highest in history. No doubt, 1973 had been a good year.

Only a few months after the replication of the 1917 start of interstate truck travel, however, earth-shattering events several thousand miles from Akron signaled the beginning of a new era in international relations: the coming global oil crisis and the worldwide energy shortage it would bring. On October 17, as a result of the Yom Kippur War between Israel and Egypt, an oil-exporting embargo was imposed by the Arab oil nations. World oil prices would soon soar, and inflation would become an international epidemic. The tire and rubber industry would never be the same again.

*Goodyear's famed Model Room in Akron. "An unhurried walk through the room," with its over 8,000 square feet of scale models of most of the company's plants around the world, "has the effect of an around-the-world junket in a Goodyear blimp."*

# Chapter XV

# 'LET'S GET BACK TO WORK'

In its first issue of 1973, the *Wingfoot Clan* gave front-page treatment to the promotions of four industrial products division executives. Robert E. Mercer, general manager of the division, was named assistant to President Pilliod. Mercer, whose 26 years of service to that point had been in industrial products, would be Goodyear's 13th president 5 years later. He was succeeded in industrial products by Westi Hansen, who had been general manager of the films and flooring division.

Donald E. Harrington, general manager of the shoe products division, succeeded Hansen. Herman B. Post, product manager for molded and extruded rubber goods, moved into the position vacated by Harrington.

Three months later, Frederick S. Myers was elected vice-president and general counsel, succeeding Arden E. Firestone who had headed the company's law department since 1960. Myers, who had been corporate secretary for a year, retained that title. He had begun his Goodyear career in 1953 on the legal staff at GAC, moved to the parent company law department in 1961, served five years as secretary and counsel at Kelly-Springfield, returned to Goodyear as assistant counsel in 1970, and was elected assistant secretary in 1971.

About 20,000 visitors — Goodyearites and their families — swarmed through corporate headquarters on Sunday, February 4, 1973, to investigate the results of a multimillion-dollar modernization program and help celebrate the 75th anniversary. Most of them were greeted by Chairman DeYoung, the moving force behind the remodeling. The changes included a terraced lobby about 100 yards east of the former lobby, Goodyear Speedramp passenger conveyor systems connecting the five floors, and contemporary offices throughout much of the complex.

The greatest alteration, though — and perhaps the most beneficial from an aesthetic view — was a new face on the entire 400-yard-long facade of the main headquarters along Market Street. Said one old-timer, recalling P. W. Litchfield's well-known aversion to investing in office facilities, "The only thing Litchfield would recognize from the outside is the address."

Among the more interesting interior changes was a new look for the famed model room. According to the unwritten corporate catechism, allowing an important first-time visitor to leave the headquarters without a tour of the model room was a sin, at least a small one.

Situated on the fifth floor of corporate headquarters, the model room, officially known as the Model Planning Room, displays scale models of most Goodyear plants around the world, each being 1/360th the size of the original. They were made initially to give management and engineers perspectives on the terrain around plants and the potentials for expansions. Emphasis is on simulation, rather than duplication; so the factory model and its surroundings present the scene at a glance, including shrubbery, pedestrians, and automobiles, all scaled to size.

Constructed of pine, the plant models have graph-paper windows and are sprayed with red lacquer undercoating to duplicate brick walls, with gray for steel. Rubber stock ready for processing is simulated by Grape-Nuts cereal painted gray. Water-tower tops are Ping-Pong balls, and metal rivets simulate air vents.

The huge model room, unique in U.S. industry, occupies 8,448 square feet, and is divided into two main sections: facilities within and those outside the United States. A large world map at the entrance contains hundreds of pinpoint bulbs that, when lighted in order, show the chronological growth of Goodyear production facilities around the world. Large murals depict the rubber plantations, the San Angelo Proving Ground, and the sprawling Goodyear complex in Luxembourg.

Plant-model making on a large scale began in 1948. William Tisen, a carpenter who had built mockups of *Corsair* fighter planes produced by the company during World War II, was assigned to create a model of the Akron facilities. He constructed more than 200 models of Goodyear's worldwide facilities before retiring in 1968 and was succeeded by Earl Bush, a veteran of tire-mold pattern design work.

A descriptive brochure available to model-room visitors points out that "an unhurried walk through the room has, in a way, the effect of an around-the-world junket in a Goodyear blimp."

## European Ambassador

Goodyear's board of directors held its first meeting outside the United States in June 1973 at Colmar-Berg, Luxembourg. As the directors toured various facilities in the huge complex, they occasionally moved in the shadow of a new traveler of the European skies, the Goodyear blimp *Europa*.

*Europa*'s primary mission is to build identity, recognition, and goodwill for Goodyear throughout Europe. Like its three sister airships in the United States, it is programmed and operated for public relations purposes. Unlike its U.S. sister ships, *Europa* operates in many different countries, frequently crossing national boundaries and facing a host of different laws and regulations with each crossing. Its first crew comprised an American cadre and Europeans who would take over after a year of training. The Europeans represented 10 countries and spoke a dozen languages.

*Europa*'s component parts were constructed in the United States and flown aboard a mini-Guppy transport plane, the *Spirit of Santa Barbara,* to Cardington, England, in December 1971. The whole airship was constructed at Cardington in a hangar maintained by the Royal Aircraft Establishment, and its maiden flight was on March 8, 1972, marking Goodyear's first airship operation outside North America and the first time any airship had flown in Britain in more than 20 years.

The British press greeted *Europa* with delight, and many newspapers published front-page pictures of its initial flight. Armed with the same Super Skytacular as the Goodyear blimps in the United States, *Europa* uses the 105-foot-long night sign to promote traffic safety, charities, tourism, education, conservation, and other public service projects.

*Europa passengers get a bird's-eye view of the famed Coliseum in Rome. Since 1972, the airship's mission has been to build recognition and goodwill for Goodyear throughout Europe as its sister ships have done so successfully throughout the U.S.*

While *Europa* was being constructed in England, its $2.5 million, 66-acre home base was being built in Italy, at Capena, 18 miles north of Rome beside the Rome-Florence Autostrada. The hangar is 250 feet long, 160 wide, and 90 high, and the huge Goodyear logos on its front and west sides are seen annually by millions of motorists on Italy's most heavily traveled highway. The base was christened in September 1972 by Mrs. Russell DeYoung.

After 10 years as board chairman, 65-year-old DeYoung retired in April 1974. He had been a Goodyearite for 46 years and had seen the company grow from an annual sales volume of $250.7 million in 1928 to $4,675.2 million in 1973. As chairman, DeYoung had steered Goodyear to the greatest growth in its 76-year history: 56 new plants were added around the world and annual sales increased by nearly $3 billion, profits by more than $100 million. At the end of '73, in anticipation of his retirement, DeYoung relinquished his position as chief executive officer. The board of directors elected President Pilliod to succeed him, and on April 4, 1974, Pilliod was elected board chairman.

## An Akron Boy

Like his two immediate predecessors as chairman, Thomas and DeYoung, Pilliod was an "Akron boy." Thomas was born and raised near the company's headquarters and its first factories; DeYoung came to Akron as a child and joined Goodyear while attending Akron University. Pilliod was born less than 15 miles from the Goodyear headquarters complex in the Akron suburb of Cuyahoga Falls. He attended grade school there and was a wrestling standout and football star at Cuyahoga Falls High School. As a high schooler, he operated a pressing machine in his father's dry cleaning shop, delivered newspapers, and worked in an ice cream store and bowling alley.

Pilliod attended Muskingum College in southeast Ohio on a semischolarship for football, but financed himself mainly by working as an auto mechanic in the summer months and in the college power plant from six to eight in the morning during the school term. He transferred to Kent State University in his sophomore year, meeting college expenses by working seven hours each afternoon and evening in a confectionery store. Pinched by the Great Depression, he decided against returning to the campus for his junior year and instead chose to get a job, accumulate funds to continue his education, and eventually study law.

As did many young Akronites of the time, Pilliod looked to Goodyear for employment. The young man had no grand designs in 1941 when he applied for a position on the training squadron; he was just looking for work. He also had approached Goodrich for work as an apprentice draftsman and Firestone for a job in the cost accounting department. He recalls that when he informed the Goodrich personnel interviewer of his potential opportunities at Firestone and Goodyear, he was told, "We would be glad to have you, but I honestly think the opportunity to join the Goodyear training squadron is by far the best of the three."

So he opted for Goodyear. After six interviews, he was assigned to the training squadron as an exception to the rule requiring its members to be college graduates. His welcome came from Charlie Jones, an ex-Marine who ran the squadron and had previously interviewed him.

"Chuck," Jones said, "you didn't make much of an impression on me, because when your name came up, I didn't know who the hell Chuck Pilliod was. But apparently you impressed the others. So you've got a job.

"But if at the end of three months I still don't know who Chuck Pilliod is," he added with matter-of-factness, "you'll be transferred out onto Market Street, because you'll be wasting your time and mine. Get your clothes and be here at 11 o'clock tonight. You'll be on the night shift."

Years later, Pilliod recalled that he worked the night shift for a year without a night off. "I enjoyed that year, " he said. "The camaraderie on the squadron was excellent. I have never been in any organization that had more spirit, pride, and drive than the squadron.

"At that time, Goodyear people — and particularly the squadron members — were convinced they were the best. The trainees had a lot of management support; Eddie Thomas, then president, used to come out for all our banquets and other functions. It was a hard-driving, hard-working outfit, full of the Goodyear spirit."

Forty years later, the same spirit was evident throughout Goodyear. In 1982, a veteran executive commented on Pilliod's approaching retirement. "We have always considered ourselves the best. Maybe we really were not, but we believed we were. We felt it and we always acted accordingly. Chuck Pilliod personified that attitude. You couldn't work with him without being convinced we would trample the competition."

## A Time for Consolidation

When the United States entered World War II, Pilliod was a supervisor for the production of fuel tanks for military use and was exempt from the draft, but he enlisted in the Army Air Corps for bomber pilot training. He later captained and flew B-29 *Superfortress* bombers over Japan from bases in India and China. Pilliod returned to the states at war's end too late to enter the fall term of college and decided to return to Goodyear on a temporary job until the next fall term.

While he was being considered for a job as squadron supervisor in Akron's Plant II, he was offered a new position in the Goodyear Export Co. on a Monday, but by Friday the budget for that job had not been approved. He went to Frank Magennis, president of Export, and said he owed it to the squadron to rejoin it.

According to Pilliod, Magennis said, "Wait a minute," then left the room to check on the budget's progress. He returned soon: "To hell with the budget; you start work with us on Monday." That began a 27-year series of international assignments that ended in 1972 when Pilliod was elected president.

He took over at a time when Goodyear's priorities were changing. He recalls that Chairman Russ DeYoung told him: "I had 10 years of growth. You're going to have 10 years of consolidation."

The new chairman, blessed with a global perspective developed in his long international career, shared his predecessor's conviction. He believed he had to "firm up our existing business in a rapidly expanding global economy, of which the United States is just a part," as he put it. "We were no longer an American company with international business. We were a true multinational with markets outside the U.S. that were growing much faster than our American markets." In Pilliod's opinion — and he expounded it throughout his career — the U.S. was just one more country in Goodyear's worldwide market.

Pilliod was succeeded as president by John H. Gerstenmaier, an executive vice-president who for three years had headed worldwide manufacturing operations; he became chief operating officer under Pilliod. Richard A. Jay, an executive VP who had directed the general products group, was elected vice-chairman of the board. Gerstenmaier had 36 years of service with Goodyear and Jay, 39.

Thomas F. Minter, who had been VP of domestic manufacturing, succeeded Gerstenmaier as head of worldwide manufacturing. Colley W. Gilchrist, manager of the tire plant in Gadsden, was named VP of domestic manufacturing, and F. Vincent Prus, corporate director of manufacturing services, moved up to VP of general products manufacturing.

These top-executive promotions once again emphasized Goodyear's unstated, but obviously firm, policy of training and developing its own management. The six men had a combined total of 208 years of service with Goodyear; Prus, the youngest at 47, had been with the company 25 years.

Longevity at less exalted levels was highlighted a few weeks after these promotions in the May 2, 1974, Akron *Wingfoot Clan*. A story on May retirees was illustrated by pictures of four married couples retiring with a combined service total of 257 years in Akron. Seven of the

eight were factory employees, and one had 32 years with the service division.

## Morocco, Malaysia . . . Mars

The financial results for 1974 underlined both the company's remarkable growth in the DeYoung decade and the Pilliod perception of increased opportunities in international markets. Sales reached $5.256 billion, an increase of $581 million in one year, making Goodyear the first company in the rubber industry to exceed sales of $5 billion. Goodyear International Corporation compiled record sales for the 20th consecutive year, and international profits amounted to 36.6 percent of the corporate total.

In mid-1974, new tire factories were inaugurated in Morocco and Malaysia. The dedication of the Malaysian plant on the outskirts of Kuala Lumpur by Prime Minister Tun Abdul Razak and Chairman Pilliod marked the official establishment of the company's joint venture with Pernas, a government group established to support Malaysian participation in local industry.

In an intracompany merger to strengthen the defense-oriented capability of Goodyear Aerospace, broaden its commercial base, and lessen its dependence on defense business which tends to vary with world conditions and government changes, the aviation products division of the parent company was joined with Aerospace in 1974. This union, of the world's largest manufacturer of tires, wheels, brakes, and fuel tanks for aircraft and 39-year-old GAC, formed a manufacturing, engineering, and marketing organization of more than 5,000 employees, which management hoped would grow through internal development or future acquisitions. Its four plants produced an array of products that included electronics, defense weapons systems, missile systems, data processors, aerial radar mapping systems, air cargo equipment, industrial and aviation wheels and brakes, and engineered fabrics.

Aerospace had been established in 1939 as Goodyear Aircraft Corp., a subsidiary that grew from 140 employees to 35,000 during World War II while handling a broader range of aircraft-building assignments than any other U.S. company. Its wartime production had included 4,008 *Corsair* fighter planes and 168 airships. In 1963, Aerospace had replaced Aircraft in the corporation's name to describe the corporation's defense and space activities.

Morris B. Jobe, who had been GAC's president since 1968, continued as CEO. Edwin M. Humphrey, general manager of the aviation products division, was named GAC executive vice-president and chief operating officer. Donald E. Hill moved from manager of the Los Angeles tire plant to VP of manufacturing, and Robert W. Clark, the division's general sales manager, became corporate director of marketing at Aerospace. Willis S. Zeigler, marketing VP at Aerospace, became VP of advanced programs and planning.

*Left, tires and braking system of the first McDonnell Douglas DC-10 tri-jet undergo simulated landing tests at Goodyear's aviation products lab, Akron, 1970. Right, Flexten cord for the belts of 2,000 automobile tires is twisted from one roll of aramid at Goodyear's Decatur, Alabama, fabric mill, 1976.*

Earlier in 1974, a material developed specifically for tire cords became part of a glamorous space project at GAC. Named Flexten, it was chosen to parachute a 2,000-pound landing craft onto Mars in 1976, and GAC would produce the lander. Flexten had been developed for belts on new radial tires, and tests showed that Flexten-belted tires had low rolling resistance and extremely light weight. In the Mars operation, this material would serve as a bridle suspending the payload from a 53-foot-diameter parachute that the spacecraft would drop.

Flexten was produced from a petroleum-derived aromatic fiber with a ringed molecular structure — a series of regularly spaced chain linkages. Using chemicals, heat, and stretching machinery in towers as tall as a nine-story building, Goodyear scientists rearranged the pattern of the molecular rings to get the stable qualities of the final product.

## Goodyear's Coldest Job

The largest single tire service contract in its history was awarded to Goodyear in early 1975 by Alaska Pipeline Service, a consortium of eight major oil companies responsible for the design, construction, and operation of the $6.4 billion, 798-mile-long Alaska Pipeline. In addition to handling all tire service work, Goodyear supplied the largest single share of tires for the project. All major tire makers furnished tires — total sales exceeded $25 million — shipping them to a Goodyear-operated tire supply center in Fairbanks.

About 100 Goodyear servicemen worked in camps along the pipeline route stretching from Prudhoe Bay on the Arctic Ocean to Valdez, the ice-free port on Alaska's southern coast. Some claimed the distinction of working in greater cold than any Goodyearites ever had. Temperatures ranged as low as minus 80 degrees Fahrenheit at stations in some parts of the state, and the chill factor sometimes dropped to as low as minus 125 degrees. In that cold, improper rubber compounds froze, breaking when kicked or given a similar blow, and parked trucks were plugged into electrically wired hitching racks to prevent oil freezing.

As Goodyear tires and tire service supported the mammoth effort to transport oil south from northern Alaska, airborne radar developed by Goodyear Aerospace searched for hidden pockets of natural gas in the Appalachian mountain range. From seven miles up, parts of Ohio, West Virginia, Kentucky, and Virginia were mapped by GAC's radar in the most extensive venture of its kind until then.

The radar used in exploration missions flown over the four-state area by Litton Industries' aero service division was developed originally for the U.S. Air Force. The so-called side-looking radar produces images of the earth's crust more distinctly than photographs. It looks sideways from the aircraft and views the terrain from an angle, providing geologists with information about faults and other formations that cannot be seen as clearly on conventional photos.

From as high as seven miles, the radar sends electronic signals downward. When these signals bounce back and are processed by electronic equipment, much as a photograph is processed, they provide maps especially suited for geological surveys.

Litton used the same GAC radar to map large areas of South America, leading to the discovery of a major new iron-ore find in Venezuela and a previously unknown 230-mile-long river in Brazil lined with huge resources of timber. The area mapped in Brazil totaled 3.28 million square miles.

*Opposite, Alaska Pipeline, 1975. Drifting snow is swept away from 50,000-gallon rubber pillow fuel storage tanks at Old Man Camp, six miles south of the Arctic Circle. About 100 Goodyearites worked along the pipeline route. Above, underwater tire reef. Weighted with concrete and dumped into the sea, bundles of tires headed for the scrap heap instead serve a useful purpose by attracting fish for feeding and spawning.*

In Brazil and Appalachia, the radar imagery recorded in usable detail an area's hydrology, geology, soil, and vegetation, all valuable in follow-up on-the-ground exploration.

A wide variety of more mundane ventures were being undertaken by Goodyear in the mid-1970s to help solve a growing ecological problem: what to do with scrap tires. By then discarded tires were being pitched on the nation's scrap heap at a rate of about 200 million annually. Goodyear tire recycling experts were beginning to develop techniques for using them as sources of energy, raw materials, and building blocks for other innovations.

In 1975, a program was initiated to burn scrap tires to supplement coal for steam generation at Goodyear's Plant I in Akron. An old, wet-bottom boiler was used to burn tires, which have a greater BTU value than coal, in the fuel mixture. About 30,000 pounds of rubber burned each day supplied 15 percent of the BTU input to the plant boiler. This innovation followed a five-year experiment at the Jackson factory, which proved that burning whole tires under controlled conditions could be done efficiently under odorless and nearly smokeless conditions. Similar experiments were conducted at the Wolverhampton factory in England.

The rubber industry's largest tire reclamation operation, at Goodyear's Akron production complex, was modernized in 1976 to reclaim and process 66 million pounds of rubber annually from more than 3 million tires. The rubber is recycled into products ranging from tires to industrial goods.

## Scrap Tires: Home for Fish

One of the more successful and widely applied recycling ideas involves underwater tire reefs that serve as fish havens. They attract fish for feeding and spawning and have no harmful effect on the environment. Within months a reef made of old tires can turn a barren sandy bottom, almost devoid of marine life, into a veritable fish community.

Tires are punched to reduce buoyancy, compacted, bundled, and weighted with concrete. The bundles are dumped into the sea at selected sites and in a short time become encrusted with marine growth — the first link in nature's food chain for fish. During the 1970s, Goodyear cooperated in more than 200 reef projects undertaken by private business, government agencies, fishing associations, boating clubs, and interested communities in the United States and around the world.

Most U.S. tire reefs were constructed on the Atlantic Coast from Ipswich Bay in Massachusetts to the tip of Florida, around the Keys, and up the long arc of the Gulf of Mexico to Corpus Christi, Texas. Some were laid on the West Coast and in a few inland lakes. Working with government agencies, Goodyear subsidiaries helped build reefs in Australia, New Zealand, Japan, Malaysia, Jamaica, Greece, and the Philippines.

Goodyear researchers also developed a floating breakwater concept for shoreline protection. Because of a rise in Lake Erie that pushed its water level four feet above an existing stone breakwater at Dunkirk, New York, 11,000 discarded tires were used in a new 1,000-foot-long breakwater there to protect marinas from storms. The Dunkirk structure, first of its kind in the Great Lakes, was engineered by the city and the Sea Grant Advisory Service and based on Goodyear-sponsored studies by the University of Rhode Island and the University of Michigan.

The breakwater was assembled from bundles of 18 tires, laced together by chains, that ride vertically in the water, supported by air trapped in the crowns. Foam and other flotation devices can be used for added buoyancy. Tests in Rhode Island showed tire breakwaters are capable of absorbing as much as 80 percent of the energy in three- and four-foot waves.

*Developed in the 1970s, breakwaters made from scrap tires are used to protect shorelines and marinas from high waves.*

One tire breakwater at Newport was credited with helping prevent a million dollars worth of damages to the International Sailboat Show when the area was hit by winds in excess of 20 knots an hour. Tire breakwaters were used to help stabilize beaches and protect them from erosion on the shores of Lake Huron and Lake Michigan.

Goodyear experts also developed scrap tire barriers for highway crash protection. The tires are cabled together to form effective cushions at freeway intersections, bridge piers, and other potential crash sites.

## Minority Strides

In the mid-1960s, President Lyndon Johnson's plans for a Great Society brought new responsibilities and some new government pressures to the nation's private sector. On the whole, business and industry responded positively to LBJ's war on poverty and his program for equal opportunity.

Cooperating with the President's Committee on Equal Employment Opportunity, Goodyear undertook a Plan for Progress in 1963 to assure equal employment opportunities and equal treatment for all employees. Chairman DeYoung endorsed this program and urged its support throughout the corporation. By the early 1970s it had gained much momentum.

Goodyear's total minority employment in the United States increased 23.7 percent from 1971 to 1974. It rose 114 percent in sales areas, 75 percent in skilled trades, and 82 percent in technical positions. In an informal report to employees via the *Wingfoot Clan* of July 31, 1975, the company disclosed its gains in minority employment during the 1972-75 period.

In 1972, Goodyear's total employment in the United States was 78,241 with minorities accounting for 10.6 percent. In 1975, despite a drop to 76,277 in total, minority employment was at 11.2 percent. The report showed women making solid gains in traditionally male job categories, in official and managerial positions previously considered only for men in a company whose character had been masculine from the outset.

In 1972, only 122 women were in the category of officials and managers. By 1975, this total had jumped to 327, an increase of 167 percent. The number of women in professional positions increased from 156 to 238 in the same period.

The year 1976 started on an upbeat note for the man who would become Goodyear's sixth board chairman seven years later. Robert E. Mercer, who had been president of the subsidiary Kelly-Springfield Tire Co. for two years, was named Goodyear vice-president for tire marketing on January 2. He had left the post of assistant to the president of Goodyear to go to Kelly-Springfield. Most of his 26-year career had been in the industrial products division. He had risen through several assignments to the division's general sales manager post in 1966 and to the position of general manager in 1968.

Mercer was succeeded as Kelly-Springfield president by Scott H. Buzby, who, like Mercer, had been assistant to the Goodyear president. Buzby had joined Goodyear in 1954 as a service representative in Pittsburgh, then served in Brazil and Colombia before moving to Australia where he was sales director and then managing director. He became vice-president of Goodyear International in 1971.

The family relationship between Goodyearites in the United States and abroad was demonstrated dramatically in February 1976 after massive earthquakes struck Guatemala, killing hundreds and leaving thousands homeless.

None of the 460 employees at Goodyear's Gran Industria de Neumaticos plant outside Guatemala City was killed, but many were injured, 30 lost their homes, and 300 more had their homes damaged. For several days, a few hundred employees and their families were forced to sleep on the ground in the factory area. As aftershocks still trembled through the Guatemalan earth, Goodyearites in Akron and Latin America were already providing aid. A series of mercy flights from Akron in Goodyear planes airlifted tents, blankets, water storage tanks, and medicine to the families of employees in the stricken country.

Responding to a request from Chairman Pilliod, Akron employees — in a dramatic display of the Goodyear "family spirit" — donated seven tons of clothing and more than $5,000 in cash. Goodyear companies in Latin America pledged $35,000 for assistance to their stricken associates, and the corporate headquarters in Akron contributed another $25,000.

## The Longest Strike

Although Goodyear sales reached a new high of $5.791 billion in 1976, a 130-day strike by the United Rubber Workers had a strong negative effect on profits and sales, making the year only "almost the best of all" for the tire and rubber industry.

Domestic tire makers had started 1976 with high hopes for all kinds of records. They counted on a resurgence of new-car sales and tire replacements for the record 11.4 million cars sold in 1973 to boost both sales and profits to new levels. Goodyear got the sales record, but its net income declined 24.5 percent from 1975. At the same time, U.S. tire manufacturers lost sales of about four million auto tires and one million truck tires, about $200 million in volume, to foreign companies. Nearly 14 million car tires were imported that year; many more would come.

The strike against the Big Four rubber companies — Goodyear, Firestone, Goodrich, and Uniroyal — was called by the URW on April 20 after talks with Firestone, the target company, had failed. Goodyear's 15 rubber manufacturing plants were struck, idling about 22,000 workers. The strike ended on August 28 with ratification of a master contract that gave the rubber workers a 36 percent increase in wages and benefits over the next three years.

In a "Now Let's Get Back to Work" message in a special edition of the *Wingfoot Clan* published on September 1, Chairman Pilliod noted: "I'm sure we are all pleased that the strike has been settled, and we can now get back to work. It's a pleasant feeling and one that should carry with it a genuine desire to work together in a harmonious and efficient atmosphere.

"At the same time," he added, "I would remind you that this is a competitive world in which we live."

In a *Clan* interview a few days earlier, he had pointed out that "the settlement will have some effect on Akron jobs." He also said the big winners in the longest strike in the history of the U.S. rubber industry were foreign companies. One of his most widely reported comments during the strike, which coincided with Michelin's establishment of manufacturing facilities in the U.S., was "God must be a Frenchman."

*Goodyear's Wingfoot trademark is stamped into a radial tire mold. The company entered the forefront of the battle against European dominance of the radial tire market in the '70s, achieving victory in '76 as the largest radial tire producer in the nation and the world.*

# Chapter XVI

# THE RADIAL WAR

Capitalizing on the radial, which it had introduced in 1948, Michelin had moved to the forefront of the European tire industry in the '50s and '60s. It took dead aim on the U.S. market as the 1970s began. The French company made its first major step into this market in 1966 when it contracted to supply Sears Roebuck with radials. Then it moved on to Detroit, selling radials to Ford for the Lincoln Continental and Mark models. By 1976, Michelin was supplying about 5 percent of Ford's tire needs and smaller amounts to General Motors, Chrysler, and American Motors.

In 1952, about 15 years before the radial tire wave edged onto U.S. shores, Michelin had about 165 dealers in the United States; they concentrated on truck tire sales. By 1976, this group had increased to about 2,000, representing more than 2,600 sales outlets, of which half sold both auto and truck tires.

Until 1971, all Michelin tires sold in the United States and Canada were imported. The company built two plants in Canada that began to produce truck tires in 1971 and auto tires in 1973. Michelin's factory at Greenville, South Carolina, the first built since its Milltown, New Jersey, factory was shut down in the 1930s, reached full capacity of about six million tires a year in 1976. It was supplied with raw materials by a new Michelin facility at Anderson, South Carolina, which had enough capacity to supply three more tire plants.

Construction of a second tire plant, in Spartanburg, South Carolina, to produce truck tires began in late 1976. That year, the French company also bought an abandoned factory from Gates Rubber Co. in Littleton, Colorado, for truck tire manufacture. Of interest to both its competitors and the URW was the absence of unions in Michelin's U.S. facilities.

Japan's leading tire and rubber manufacturer, Bridgestone Tire Co., Ltd., also made inroads into the U.S. market in the 1960s and 1970s and greatly increased its share of the world's tire market. Like many Japanese manufacturers, it had conducted a strong export drive in the 1950s. In the '60s, it built tire factories in Singapore and Thailand and established the Bridgestone Tire Company of America, Inc., at Torrance, California, to handle most of its increasing exports to the United States. It opened nine new tire factories in Japan from 1960 to 1976, greatly augmenting its export capability.

By the late 1970s, Bridgestone was the sixth largest tire company in the world with 16 plants worldwide and sales approaching $2 billion. Industry observers were beginning to believe Pilliod's often-expressed opinion that in the '80s a Big Three of Goodyear, Michelin, and Bridgestone would stand unchallenged at the top of the global tire industry.

The radial began its dominance of the U.S. market in 1976 when it took over 45 percent of U.S. tire sales. By then, Goodyear had emerged as the largest radial tire producer in both the nation and the world. The two largest 100 percent radial passenger tire plants in the United States were Goodyear facilities at Union City, Tennessee, and Fayetteville, North Carolina. Further, the company announced a $69 million expansion of its Gadsden, Alabama, factory for their manufacture.

## CEO's Strong Influence

The effect of Goodyear's chief executives on the company's direction and progress is evident throughout its history. Frank Seiberling had founded Goodyear as an aggressive entrepreneurial organization and guided it to the top of the world's tire industry. E. G. Wilmer helped shore up its finances after the near bankruptcy of 1921. Paul Litchfield assured its leadership in technology, accelerated its diversification, provided the early impetus for its international expansion, and, perhaps more than anyone, gave Goodyear its character and personality.

E. J. Thomas, who personified the upbeat Goodyear Spirit, infused marketing and communications with a new vitality, pushed across new foreign frontiers, and evoked a corporate-wide attitude of confidence and enthusiasm.

Russell DeYoung led the company through its greatest growth, widened its geographic horizons, and considerably strengthened its international product and marketing operations. Chuck Pilliod, in addition to consolidating Goodyear's global leadership, converted it from a U.S. corporation with foreign subsidiaries to a true multinational, drove it to a preeminent position in the Radial Age, and both restored and reinforced its technological and innovative capabilities.

On assuming the corporate presidency in 1972 — and with strong support from DeYoung — Pilliod began a gradual but firm push to improve and unify Goodyear technology on a global basis, to shape an international corporate perspective, and to outstrip competition in research and development, particularly in radial tires. Those efforts attained full speed in the early years of his chairmanship, leaving no doubt where he wanted the company to go and how he visualized its getting there. Tough minded, forceful, and a constant articulator of major policies, he laid out a well-marked path.

Blessed with natural leadership qualities, Pilliod strengthened and refined them during his rise to Goodyear's top job. When he left Brazil in 1963, after three years as commercial manager and four as managing director, he was an almost legendary figure among employees and dealers. He had built an exceptionally strong and loyal dealer organization in a highly competitive market, and Goodyear do Brasil was well on its way to becoming the company's largest foreign subsidiary.

A story is still told there about a conference of Goodyear salesmen and dealers, held in a provincial city. The meeting went on into the late hours of Saturday night, and when it came time to conclude, Pilliod said: "OK. Let's knock off. I'll see you all at eight o'clock Mass tomorrow morning."

Catholic or not, everybody was at Mass. Whether legend or truth, the story both represents and reflects something of Pilliod's leadership style.

Just as friendly and informal as forceful, Pilliod eschewed pomp and the trappings of executive power, obviously preferring Chuck to Mister. His manner was uniform in dealings with all: high-level executives, Washington politicians, clerks, secretaries, college students, and airplane mechanics. As a long-time pilot, he gave as much deference to the mechanics as to anyone. During his stint as chairman, he often copiloted company jets on business trips, reluctantly giving this up at 62, the compulsory retirement age for Goodyear pilots.

Like DeYoung and Thomas, Pilliod was highly competitive in business and sports. A physical-fitness enthusiast from his college days on, at 64 he still arose at 5:00 A.M. for nearly an hour of exercise, usually arriving at his office by 7:30.

Taking over amid many new outside confrontations for business, Pilliod had to go public much more than the chairmen who preceded him. Goodyear needed to respond openly to a multitude of new pressures — among then consumerism, environmentalism, equal opportunity, wage and price constraints, burdensome government interference — and Pilliod brought enthusiasm and great energy with what seemed to be a natural aptitude in meeting them all head-on.

His travel schedule, in fact, resembled an airline pilot's; rarely did he go a week without a quick visit to Washington or New York, a month without brief stays in foreign areas.

*Chairman Pilliod speaking informally at the dedication of the gas centrifuge plant in Akron, 1976. Pilliod, more than any Goodyear chairman before him, faced a need to go public on a variety of issues and was well equipped for the job.*

Despite his long absences from Akron in long-term overseas assignments, Pilliod remained an Akron lad. On assuming the top job at corporate headquarters, he became deeply involved in the progress and well-being of the local community. He was the first chief executive of a major Akron employer to serve as campaign chairman for the annual United Way drive.

In 1978, he then set up — and sold the idea to Akron — a system whereby other chief executives of major companies would follow each other in this job. In the same year, he also served as chairman of the national U.S. Savings Bond campaign, plane hopping around the nation to kick off regional or local campaigns in 23 cities.

Pilliod believed industry owed to the system of democratic capitalism more than the operation of a sound business that met public needs while being a fair employer and a decent corporate citizen. He also felt the marketing system got insufficient exposure in the halls of higher education, compared with other systems based on antithetical philosophies, and that industry had a responsibility to see this situation improved.

So in 1973, he initiated a program in which Goodyear established "chairs of free enterprise" at the University of Akron and Kent State University. The objectives of courses conducted from these chairs were to foster entrepreneurial spirit, combat economic illiteracy, provide a forum for the discussion of business concepts and the business enterprise system, and engender greater appreciation of careers in business and industry.

The courses quickly became popular and have been highly visible in the business school curricula at both universities. Generally the occupants of the chairs accept a one-year assignment. But Felix Mansager, retired chairman of the Hoover Co., held the position at Akron University five years. He developed a "Students In Free Enterprise" program in which, near the end of their courses, the students visit Akron area high schools and grade schools to explain the free enterprise system to pupils.

Although disclaiming any more than an average social conscience and despite devoting 10 hours a day to his direct Goodyear responsibilities, Pilliod probably was as involved as any Akron resident — often behind-the-scenes — in community programs. His secretary during his last four years as chairman estimated that about a third of his incoming calls concerned national and local affairs.

## An Eye on Tomorrow

Never was there a doubt, though, that Pilliod's eye was constantly on the corporate bottom line of today and tomorrow. Despite his outside-the-office activities, he ran the company. To the same degree as Litchfield, Thomas, and DeYoung, perhaps even more, he knew what was going on throughout Goodyear. Indeed, the impetus for much of what *was* going on came from the chairman's office.

On assuming the corporate presidency in 1972, Pilliod laid out a plan for Goodyear's progress during the next decade, the end of which would bring his retirement. He enunciated the plan frequently as president and even more frequently after becoming chairman and chief executive in 1974. A favorite platform for restating and explaining his program was the quarterly meeting of about 400 U.S. and Canada managers held in the Goodyear Theater in Akron. Plant managers throughout the U.S. and executive committees overseas were kept informed of what went on at the meetings through tape recordings and written synopses.

Upon his retirement, Pilloid remarked with some pride that the plan had never been changed and had been successfully followed to its ultimate objectives. That had required rapid and large-scale conversion to radial tire manufacturing and the rebuilding and restructuring of Goodyear's American tire factories.

The restructuring would be expensive. Other American tire manufacturers were unwilling or unable to commit the necessary funds. Some, realizing the change would come and that it was a time to commit investments accordingly, made the decision to invest in diversification outside the tire business. Others chose to convert to radial production, but slowly and without major investments. Still others simply did not have the resources to make the change.

Goodyear's decision was to go all out to develop the most advanced radial manufacturing processes and establish itself firmly as number one in radial tire production and technology.

Some rubber industry observers called the plan bold and daring. Pilliod, who had quickly built his own reputation for boldness and daring, disagreed. Looking back after retirement, he described his perception of the times.

> There was nowhere else to go. To continue along the old lines would have been very difficult, as some of our competitors found; and it would have meant capitulation to foreign radial tire competition, mainly Michelin and Bridgestone.

The financial objectives of the Pilliod plan were to achieve a 5 percent return on sales and 15 percent on investment while reducing debt equity to 35 percent. "Once we had reached those goals," Pilliod explained, "we would consider diversification, but not until then."

Underlying the plan was a strong faith in the future of the tire industry. Pilliod often used a favorite example to explain that faith.

> Kids at the age of 15 years, 11 months, 29 days, and 23 hours tick off the minutes to their 16th birthday and then say, "Dad, I want my temps [temporary drivers license]." Then when they get a job, their first ambition usually is to save money to get "wheels." So long as that goes on — and I don't see it changing — the tire business is going to be there and be good.
>
> So if you're going to be in the tire business, you'd better be the best, because in our free enterprise economy there always will be some at the top who are leaders and make money, some in the middle who just plug along, and some at the bottom who lose money . . . We're going to be number one in the tire business as we have in the past; and we'll continue to be.

Pilliod had other reasons for commitment to tire industry leadership — dedication to advanced tire technology and responsibility to the dealer organizations.

*Goodyear radial-ply tire expert Don Roberts shows the bias angle of plies in the conventional tire cutaway at left as compared with the horizontal angle in the radial-ply cutaway and the less extreme angle of the four-layer belt, 1967. At right, Goodyear's Power Cushion Radial-Ply tire. The key to the radial tire's ability to produce up to 100% greater tread life than the conventional tire is its stiff belt, which acts like a hoop.*

If we allowed our U.S. tire operations to fall behind, we would have great difficulty in providing research and development backup to our foreign operations. So, we just had to keep our U.S. base business — tires — in top form. We also had a big responsibility to our dealers. In a sense, they are our partners. If we started to drop out of the tire business, we would be letting them down, all over the world. And their goodwill represents a huge asset. Building or rebuilding that kind of goodwill, in any business, would be very expensive.

So, in the early '70s, it really was easy to say we were not going to diversify outside the tire industry. We resolved to build our world tire leadership, increase our efforts in that business, and strengthen our capabilities as the leader in tire technology, manufacturing processes, product design, and product quality.

Pilliod had hoped to reach his financial objectives in five years, but two recessions got in the way. In 1982, the goal of less than 35 percent debt equity was reached and the 15 percent return on investment was near, but the 5 percent return on sales was still 2 percentage points away.

A major source of pride to Pilliod as he looked back over his nine years as chief executive officer was Goodyear's avoidance of crash personnel reductions as recessions moved in. On taking over as president in 1972, he put a block on large-scale hiring. So, when many companies were forced to slash personnel rosters in the late 1970s and early '80s, Goodyear's rolls already had been trimmed. In 1982, the company's total worldwide employment was nearly 15 percent below 1974, although sales had nearly doubled.

Two years after Pilliod's election as board chairman, Goodyear was moving fast toward two of his major objectives: a quantum leap forward in its technology and a globalization of its perspectives and operations. The two were tied closely together. As tire markets overlapped national boundaries and the global market became more cohesive, he constantly emphasized the need for transnational technology and the challenge of rapid advances in tire technology in Europe and Japan.

Throughout his many years in international operations, Pilliod had watched foreign technology — first in industrial rubber products, then in tires — edge closer to technology in the United States. He later recalled losing ground in industrial products markets abroad because U.S.-based technology, on which Goodyear's foreign subsidiaries were then greatly dependent, was not progressing as rapidly as technology in some foreign countries.

One of his major ambitions, expressed frequently in management meetings, was to enable Goodyear to "get a leg up" in technology on a global basis. He insisted on cross-fertilization between research and development operations in the United States and in foreign countries, particularly between R&D headquarters in Akron and the International Tire Technical Center in Luxembourg.

After battling the radial tire in United Kingdom markets for three years, Pilliod had returned to the United States in 1966 convinced that this tire was indeed the tire for today, not just the tire of tomorrow as many in the U.S. tire industry viewed it. As director of operations, vice-president, and president of Goodyear International, he had pushed hard for advanced radial technology and greater radial capacity in Europe. When he assumed the corporate presidency in 1972, he broadened this push to include the United States, and at first he met some resistance.

"Although our development people knew the radial tire was there to stay in Europe, they still were not sure it was the tire for the states," he said in retrospect.

> I sent people to Europe so they could convince themselves we must go radial in the states. But they came back and said it wouldn't go here; it was too expensive and people wouldn't buy it.
>
> I couldn't get figures from our market research people to support radial expansion in the U.S. So that's when I said, "Look. I hear you talking, but you're wrong. And we're going flat out to install capacity for 100,000 radial tires a day immediately."
>
> And really, we never stopped. In those days at Monday morning meetings with production, machine design engineering, and development, we set up crews to search the world over to find the latest and most-efficient radial equipment. Then we tried to design something better. We figured if we were to be producing radials for the next 20 or 40 years, we should have equipment that would give us a leg up.

*Air traffic controllers monitor Staran, developed by Aerospace in 1971 and called the world's fastest computer.*

At that time, Pilliod hastened efforts to mesh U.S. and international development operations. "It was obvious," he said, "that we had to use all the best brains we had on both sides of the Atlantic. But it took some effort to convince all our R&D elements in the U.S. to take a global approach."

That effort had gained corporatewide support by the mid-1970s, and in 1976, the research and development division was reorganized for closer alignment with corporate objectives and greater emphasis on advanced concepts and high technology.

## R & D Gets Top Rank

In June 1976, Thomas F. Minter became Goodyear's first executive vice-president for research and development, including machine design, a post to provide single direction of these previously separate functions. Minter was assigned to the corporate policy committee and the board of directors, giving R&D a representation at the highest management level. In 1974, he had been elected executive vice-president of manufacturing and a director, and before that he had served nearly three years as VP of domestic manufacturing.

Several other top-management moves were made at that time. Richard A. Jay, vice-chairman, was named president of the general products division and continued to have the responsibility for Goodyear Aerospace and Motor Wheel Corp. Robert E. Mercer, formerly vice-president of tire marketing, was elected an executive VP of the corporation and named president of the tire division. He succeeded Charles A. Eaves, Jr., who would soon retire.

Four other promotions were involved in the management restructuring. Colley W. Gilchrist, VP of domestic manufacturing, was elected executive VP of corporate manufacturing, succeeding Minter. Tom H. Barrett, production director, Goodyear-Luxembourg, was elected VP of domestic manufacturing.

Under Minter's direction, the research and development division was reorganized and Robert E. Workman was elected to the new position of vice-president with responsibility for general products development. At the same time, Westi Hansen returned from a three-year assignment as managing director in the United

Kingdom to assume the position of vice-president of the general products division.

Goodyear Aerospace's prospects were not bright at the start of the 1970s. The national economy was slumping, competition for defense contracts was intense, and the aerospace industry was flat.

Even as the Vietnam War neared its end and the nation's immediate military requirements lessened, the pace of GAC's technological advances quickened. By the end of the decade, for example, it was the nation's leading contractor for gas centrifuges to enrich uranium for nuclear power.

Between 1968 and the end of the Vietnam conflict in 1973, defense spending had been reduced by one third in real terms, according to an article in the *Aerospace Clan*, GAC's employee publication. Declining employment at the subsidiary's Akron complex caused concern among management and employees and in the community.

The mood of those working in the shadow of the huge Airdock was far from cheerful, but horizons took on a brighter hue in 1975. Perhaps the announcement that year of GAC's successful design and manufacture of gas centrifuge units for uranium enrichment was a good omen, as some GAC veterans had said. In any event, business picked up in 1976 and attained record post-World War II levels each year through the remainder of the '70s.

## GAC: Growing Capabilities

GAC ended 12 years of production of Subroc antisubmarine missiles in 1972, but the Navy awarded the company a $26.3 million contract for engineering development of Captor (encapsulated torpedo), a new antisubmarine weapons system.

Throughout the 1960s, Aerospace scientists worked continually on the Staran computer system for advanced electronics research. One result was the Staran IV, called the world's fastest computer, for air traffic control; it could perform 40 million mathematical operations per second. The National Society of Professional Engineers named Staran one of the 10 outstanding engineering achievements of 1971.

Staran, Captor, the 98-foot-long *Viking* parachute that helped soften the *Viking I* lander's arrival on the surface of Mars, and the airborne side-looking radar were four of the most glamorous GAC products of the 1970s and generated new attention to the company's growing capability in high technology. Many other products came from GAC's Akron, Arizona, and Berea facilities in that decade; few, though, were made-to-order grist for the publicity mills.

Some, but by no means all, were:

- Cargo containers, windshields, canopies, and radar domes for the McDonnell Douglas F-4 *Phantom* multipurpose fighter;
- Lightweight armor for Air Force crews;
- Bondolite flooring for the Douglas DC-10 and the Boeing 747B;
- Air brakes for the Air Force's McDonnell Douglas F-15 fighter;
- Rubber sonar domes for Navy destroyers;
- A variety of flight simulators for the Air Force;
- Antiskid systems for Air Force T-39 jets built by Rockwell International;
- Fuel tanks for the Bell 214ST helicopter;
- An automatic braking system for the Lockheed L-1011 *Tristar* jetliner;
- Brakes for the Navy's F-14 *Tomcat* carrier-based aircraft.

Several notable large contracts were awarded to Goodyear during the 1970s.

- $87 million for side-looking airborne radar for West Germany's RF-4E reconnaissance aircraft
- $33.2 million from Martin Marietta for full-scale development of a radar guidance system for the Pershing II missile system
- Eight contracts from the Navy to produce MK-48 torpedo warheads, electronic assemblies, exploder, and associated parts

The largest single contract since World War II came from the Navy in 1979, calling for additional production of the Captor mine system.

GAC's 1976 increases reflected in sales, 4 percent; net income, 79 percent; new orders, 12 percent; backlog, 45 percent. A veteran employee remarked, "This must be the happiest time of Jobe's 35 years in GAC."

That certainly was possible, because Aerospace obviously was out of the doldrums, and Morris J. Jobe, its president since 1968, confidently predicted 1977 would be an even better year. It was, and so was 1978, but even greater satisfactions lay ahead.

Goodyear Aerospace signed a $90 million contract in 1979 with the U.S. Department of Energy for centrifuge production. Goodyear had become the first industrial company to qualify a gas centrifuge and in 1979 was the top producer in the centrifuge program. The company's investment by that time amounted to more than $20 million. For its part, the government had invested nearly $400 million in the centrifuge research and development program from 1960 to 1979.

Dedicating the new $26 million centrifuge building at the GAC complex on July 26, 1981, Chairman Pilliod said:

> The guy who started this project is Morrie Jobe, who retired a year ago as president of Goodyear Aerospace. Morrie wasn't at all retiring in 1972 when he proposed this project. In fact he was downright positive and aggressive.
>
> Morrie made a presentation on the role of nuclear energy in our country's future. He expressed the firm conviction that Goodyear could participate in the national nuclear program through production and sale of nuclear enrichment equipment — namely centrifuges. At that time, no centrifuge equipment had been mass-produced in this country. So if Goodyear should undertake such a program, it would be on a pioneering basis . . . The sums involved were high — and so were the risks, since we would be in a totally new field of endeavor for Goodyear. But, Morrie was convinced Aerospace had the technology to do the job, and he convinced us.

Without knowing it, Goodyear probably had entered the nuclear age 39 years before starting its centrifuge operation. In reminiscences published in 1982 during the observance of the 40th anniversary of nuclear fission, the widow of physicist Enrico Fermi indicated the first nuclear chain reaction apparently took place in a "cube-shaped balloon" built by Goodyear. Fermi, physicist Arthur Compton, and others produced the first nuclear chain reaction at the University of Chicago in December 1942 as part of the Manhattan Project.

They needed a balloon, according to Mrs. Fermi, to contain the atomic pile and trap air escaping from it. Goodyear was not advised of the details of the project or the end use of the balloon. They were only asked to produce it.

The cube-shaped balloon, however, was the subject of many jokes by the company's engineers, specialists in designing aerodynamically shaped blimps and other lighter-than-air craft. They knew it would never fly, but then, it was never intended to.

By mid-1977, the drive for greater innovation had gained headlong momentum and high visibility throughout the Goodyear world — scarcely more than a year after Chuck Pilliod had started a major push in that direction.

## Innovation and the Tiempo

Early in 1976, a consulting firm had been retained to examine all facets of company operations affecting technological innovation. Its findings brought on several structural changes, including these major ones:

- The consolidation of research and development functions under Executive Vice-President Tom Minter to give central direction to the corporate technical effort;
- A strengthening of the general products development function under Vice-President Robert Workman;
- The creation of a project management system under Carl Snyder, associate director of research who became vice-president of research in 1979;
- The establishment of an innovation policy committee chaired by Wendell Minor, vice-president, North American tire subsidiaries.

In a special Innovation Edition of the *Wingfoot Clan*, on July 28, 1977, Pilliod challenged all employees to help the company innovate.

"It is time to make sure every Goodyear employee knows how important innovation is," he wrote. "We want every contribution our employees can make. Just being the biggest is not enough; you must also be the best. Our ability to come up with new products, processes, and services is vital if Goodyear is to maintain its industry leadership. This applies not only to those involved in product design and manufacturing, but to all personnel from the lowest to the highest echelons."

Underlining its commitment to innovation, Goodyear introduced a revolutionary gas-saving tire that same week. Although not a marketing success, it was a big step toward the gas-efficient, high-inflation radial tires of today. Several of its characteristics have been incorporated in two of Goodyear's most successful contemporary tires, the Tiempo and the Arriva.

The new tire featured a lower profile than radials of that time and an elliptically shaped sidewall that formed a curve down to where tire meets wheel rim. In conventional tires, including bias ply, bias belted, and radials, the sidewalls are straight in the area immediately above the beads that grip the wheel. The built-in continuous curve of the elliptic tire helped it flex and absorb road bumps with no decrease in control and handling, despite the tire's high inflation pressure.

As tire engineers had known for years, high inflation reduces a tire's rolling resistance and thus improves fuel economy. Until the elliptic tire, though, high inflation had caused harsh rides and poor handling, and passenger radials rode most comfortably with an air pressure of 24 to 28 pounds per square inch (psi). With the elliptic, this became 35 psi.

The elliptic tire required a wheel with a rim design of slightly lower flanges than traditional rims. This worked against its acceptance by cost-conscious Detroit, which would have to retool for the new flanges. As introduced, therefore, the elliptic tire never reached the marketplace, but its tread-design variations and some of its characteristics, including the 35 psi, soon became part of the modern radial.

Seven weeks after introducing the elliptic tire, Goodyear announced a new all-season tire to help end winter tire changeovers. It was designed for use on dry pavement or in rain or snow and was called the Tiempo.

In announcing the Tiempo, Robert E. Mercer, executive vice-president and president of the tire division, explained that its winter traction was provided by shoulder elements extending half an inch into the center of the tread and by specially formulated rubber compounds. Mercer predicted the Tiempo would account for more than half of Goodyear's winter radial tire sales in the coming year and up to 33 percent of total radial sales in the nation's snow belt region.

It exceeded these expectations, and more. The Tiempo enjoyed greater initial market success than any tire in the company's history.

## The Lawton Plan

In 1973 when the United Rubber Workers struck the tire industry following contract negotiations, Goodyear was one of very few rubber companies without a nonunion tire plant. Thus, it was more adversely affected by the strike than competitors with some nonunion factories.

As the URW had a long history of not settling without a strike, Goodyear's management decided some sort of offset was necessary. Pilliod determined to consider a new approach to labor relations, beginning with a new tire plant.

*The Tiempo radial, Goodyear's most successful new tire. Deep shoulder lugs, unique tread design, and special rubber compounds give winter traction and a smooth ride year-round.*

As a first step, the industrial relations division was reorganized, with individual directors assigned to union and nonunion operations. This permitted direct alignment of the unionized plants with the terms of existing labor contracts and provided flexibility to explore new fields in labor relations.

Clyde R. Boose, who had served for more than three decades in Goodyear's industrial relations, remained as industrial relations director for the organized operations. Philip E. Ensor, a widely recognized expert in the "people motivation" aspects of industrial relations, was brought in as director of nonunion operations. One of Ensor's first assignments was to assist in selection of management personnel for the new plant.

In early 1977, plans had been announced for the plant. It was to be at Lawton, Oklahoma, and would represent the largest initial factory investment in Goodyear's history. It would be the most highly automated tire manufacturing facility in the world and would employ about 1,400 workers in a 1.449 million-square-foot facility on 500 acres. Ground was broken on June 14 by Goodyear President John Gerstenmaier and Oklahoma's Governor David Boren; the *America* floated overhead.

The Lawton operation was planned and operated from the start to "produce the best, most-modern radial tire in the world," according to Gerstenmaier. The key ingredients were to be the latest in tire-manufacturing technology and well-trained people with mutual goals working in an atmosphere of mutual respect.

Lawton was Goodyear's first operation where factory management had absolute control of all materials and processes. Every component was — and is — under constant computer monitoring, from the raw materials receiving area to the finished products coming off the line, and anything is retrievable at any stage. The entire process is controlled by a master computer. On the human side, "people programs" were established at the outset to involve employees in every aspect of the work environment.

Managers were chosen for their proven "people orientation" even at the expense of experience or education. The aim was a highly productive work force built on mutual understanding and consideration of both employee and company needs.

The program was a success from the start, putting Goodyear among the leaders in enlightened industrial relations. By 1981, when the Lawton plant reached full capacity, turnover of manufacturing employees was less than 1 percent; it dropped to 1/3 of 1 percent by early 1983. The accident incident rate, which was 10.4 percent in 1981, was 2 percent at the start of 1983. In the 1981-83 period, controllable absenteeism declined from 1.5 to 1 percent. With the extension of the program to unionized plants, Goodyear entered a new era of employee-management relations which not only fostered harmonious industrial relations but also was instrumental in attainment of productivity and product quality goals.

Lessons learned at the Lawton factory were applied gradually at other and older plants, with excellent results. A prime example is Motor Wheel's plant at Lansing, Michigan, a city so ravaged by the 1980-82 auto industry recession that its unemployment in late '82 was 12 percent, and more Motor Wheel employees there were laid off than were working.

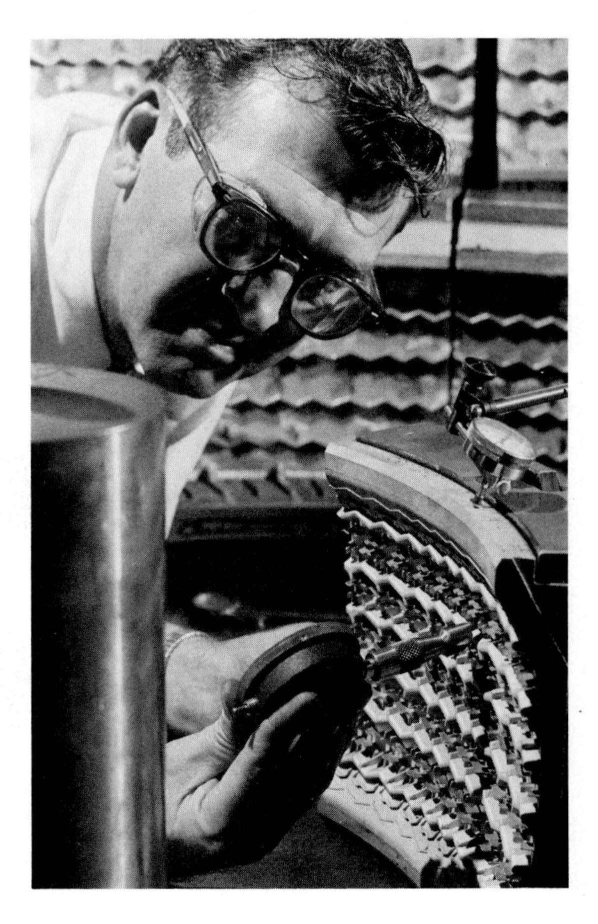

Application of Lawton Plan techniques at Lansing brought productive management-union cooperation and joint problem solving to an operation that until then had been beset by acute adversarial relationships between labor and management. Motor Wheel's employee involvement plan, an adaptation of the Lawton Plan, featured supervisor-shop steward training, jointly led and jointly attended.

Joint union-management problem-solving teams were set up, and a task force went through the factory giving organizationwide instruction on the factory floor in how to get all employees at all levels involved in working together to solve problems. The result was a drastic turnaround in industrial relations: from extreme discord to fruitful harmony between employees and factory management.

Concentrating on the human factor, the Lawton Plan — even when applied to older plants — proved the value of organized interpersonal relations and worker participation in problem solving. It was also a new way of paying more than lip service to the long-standing Goodyear motto, "People Are Our Most Important Asset." As some pointed out, perhaps it was not really new, but just a return to the kind of close personal management-employee relationships P. W. Litchfield found when he came to the tiny company in 1900.

If any skepticism had existed about Goodyear's continuing commitment to innovation and its faith in the future of the automotive industry, it was dissipated in August 1978 when Chairman Pilliod announced plans for the Goodyear Technical Center in Akron. This project would revitalize a faded, 3,000-acre area less than a mile from corporate headquarters and "reaffirm our future in Akron and our commitment to innovation," he said.

*Left, checking tread grooves in radial tire mold. Below, top, technicians at San Angelo, Tex., proving ground prepare to drive elliptic tires over slicing probes to test resistance to penetration. Bottom, "green" or uncured tire inspection, Lawton, Okla. Right, freshly-cured Tiempo radial tire passes through one of a series of inspection stations at Union City, Tenn.*

## 'Bold Step'

Pilliod called the center "a bold and exciting step made in the spirit that propelled a fledgling enterprise, born 80 years ago this month [August 29, 1898] into the world's largest tire and rubber company." He said:

> Our ability to create new products, processes, and services is vital if Goodyear is to maintain its industry leadership.
>
> The Technical Center — and an ancillary test center and test track — will give our engineering and technical personnel the environment, facilities, and equipment to meet the challenge. We look forward to the future with optimism and confidence that "The Greatest Name in Rubber" will reach innovative heights as the company approaches the threshold of the 21st century.

With an eventual total investment of a quarter-billion dollars and a construction timetable of three to five years, the Goodyear Technical Center is believed to be the largest single private investment in Akron history. Its centerpiece is the converted Plant II tire factory, a 62-year-old landmark idle since the spring of 1978 when Goodyear discontinued auto tire production in Akron. The 700,000-square-foot building will employ about 1,200 people after conversion to offices, laboratories, and test, engineering, and design facilities, plus a miniproduction setup with a capacity of up to 1,200 experimental and specialized tires a day. A one-mile performance and handling test track was built several blocks from the center's main building.

Management admitted the center could have been located in other areas of the U.S. However, it felt a responsibility to help offset the employment losses of recent plant closings in Akron and wanted to revitalize that section of town. The city and state governments supported the project, covering a portion of the cost of clearing the area and relocating roads, bridges, and parks. The project spearheaded the long overdue rehabilitation of part of Akron's east side.

The five-story main entrance facade is virtually encased by glass, which provides a sky-view ceiling for the lobby and product display and reception areas. The office area, on floors four and five, is built around a glass-enclosed atrium. Space on the first, second, and third floors has been converted into a miniaturized manufacturing operation where engineers and technicians can follow their ideas from the drawing board to finished experimental products.

"It's certainly a far cry from Paul Litchfield's tiny experimental department established 70 years ago," said Frederick J. Kovac, who in 1980 succeeded Robert B. Knill as vice-president, tire development. "But the determination of those days to be number one through innovation and advancing technology remains the same."

## Push for New Ideas

Management continued to press for innovation companywide throughout the Pilliod Era. A plan for employee recognition and awards for new ideas, to be conducted in conjunction with the long-standing Goodyear Suggestion Program, was announced in August 1979. The program, established in 1912, evaluates cost-saving ideas or recommendations to improve health, safety, or working conditions. Throughout the corporation, 4,075 suggestions were submitted in 1979, of which 1,103 were adopted. They resulted in annual savings of $1,012,340 and a total of $125,395 in award payments based on minimum cash awards ranging from $15 up to 10 percent of the gross annual savings achieved by the suggestion.

The New Ideas Program, established to recognize truly inventive, creative, or innovative achievements, offered awards ranging from gifts of merchandise to $50,000, the larger being for special ideas of value to the corporation. For ideas of value on the divisional level, special awards ranged from $500 to $5,000. The program's theme was "Innovation Starts With You," and when in 1980 it had gained full momentum, it had attracted 4,640 ideas, of which 21 percent was implemented at a cost savings of $1,336,855. Awards with a total value of $72,300 were given to employees with accepted ideas.

In concert with the federal government and other major rubber companies, Goodyear applied its innovation capability to a late-'70s effort to make rubber from a North American desert bush called guayule. The bush produces a commercially feasible source of rubber. Mexican guayule, called the "rubber orphan of the wastelands," had supplied 50 percent of all U.S. rubber and 10 percent of the world's requirements in 1910. However, falling rubber

*Above, the amount of rubber available from a guayule bush is determined using nuclear magnetic resonance technology. Below, visitors to Goodyear's museum in Akron are shown a guayule shrub and a tire made with rubber from guayule.*

prices, greatly increased production of tree-grown rubber in the Far East, and the Great Depression had combined to end its use.

An Army major named Dwight D. Eisenhower recommended in the 1930s that the government nurture guayule as a second source of rubber, imported tree-grown being the first. His suggestion was to no avail.

With World War II and the loss of Southeast Asia — source of 90 percent of the Allies' rubber — Washington took action. More than 30,000 acres were planted with guayule at 13 sites in three states. They supplied raw product to the wartime Emergency Rubber Project, which, at its peak, turned out 15 tons of rubber a day from California-based factories in Salinas and Bakersfield. Four mills in Mexico processed 180 tons of guayule shrub every day, but the program was aborted following start-up of the wartime synthetic rubber industry. By 1950, the Emergency Rubber Project was only a memory.

In the mid-1970s, interest in guayule again picked up when the international oil crisis caused uncertainty about the price and availability of petroleum feedstocks and doubts that there would be enough tree-grown rubber to satisfy demand during the '80s.

Following a call by the National Academy of Scientists for "a national commitment to guayule research and development," Congress passed the Native Latex Commercialization and Economic Development Act of 1978 and allocated $40 million to explore the desert plant's economic potential. The major tire companies swung into action. Goodyear's program was directed by James D. D'Ianni, director of research. The rubber industry began cultivating millions of guayule seedlings on experimental farms in Arizona and Texas.

Goodyear — the world's biggest user of both natural and synthetic rubber — went a step further. Engineers in Akron produced experimental guayule auto tires that passed the Department of Energy's high-speed and endurance tests. They also made a 10-foot-tall, 5,000-pound guayule earth-mover tire. Nearly a ton of guayule was used in the big tire, part of a 4,500-pound supply the company had been accumulating for a year.

In September 1980, Goodyear formed a joint project with the eventual goal of a domestic natural rubber industry based on guayule. Coparticipants were three Indian tribes and two Arizona crop-research organizations, the Agri-Business Research Corporation of Scottsdale, Arizona, and the Center for Arid and Tropical New Crop Applied Sciences and Technology of Arizona State University at Tempe.

The project, slowed by economics and more oil imports at better prices than in the 1970s, has moved slowly. Economics is a retarding factor because guayule latex cannot be drawn directly from the standing bushes, as it can from the Hevea tree, which produces nearly all the world's natural rubber. The entire guayule plant — roots, stems, branches, leaves, even the dirt and stones around it — must be torn from the earth for processing. So although guayule rubber can be of high quality, processing is expensive. Goodyear experts feel that American industry can succeed with it only with government support, with greater need for it, and with production that allows the pricing of guayule to be competitive with tree-grown and synthetic rubber.

*Robert E. Mercer (left) and Charles J. Pilliod, Jr., Goodyear's sixth and fifth chairmen, in the Model Room.*

## Ascension in the Ranks

In May of 1978, Robert E. Mercer took the penultimate step toward leadership of the world's largest tire and rubber company. He was elected Goodyear's 13th president, succeeding John Gerstenmaier. Elected to the newly created position of vice-chairman and chief operating officer, Gerstenmaier was given responsibility for corporate staff functions serving both the U.S. and international operations including production, research and development, personnel, purchasing, and quality control. Mercer's new responsibilities included the United States, the tire and general product divisions, and the company's U.S. and Canadian subsidiaries.

Gerstenmaier, who joined Goodyear as a trainee in 1938, had advanced through the production organization, rising to vice-president of domestic manufacturing in 1967 and executive VP for worldwide production four years later.

Mercer, a Goodyearite since 1947, was president and chief executive of Kelly-Springfield from late 1973 until early 1976. Earlier he was a key figure in the industrial products division, its general manager from 1968 until 1972 when he was made assistant to Pilliod, who was then president. He became a corporate officer in 1976 when elected to the vice-presidency of tire marketing.

Scott H. Buzby, formerly president of Kelly-Springfield, was elected to succeed Mercer as executive VP of the corporation and was named president of the U.S. tire division. Buzby had moved up through the ranks overseas in

Goodyear International and was elected its VP for sales in 1971. He was named assistant to the president of Goodyear in 1974. Succeeding him as president of Kelly-Springfield was Albert H. Shafer, Goodyear's VP of replacement tire sales.

Five months later, in October, another future Goodyear president moved up to an executive vice-presidency. Tom H. Barrett was elected to that post in corporate manufacturing, succeeding Colley W. Gilchrist, a 42-year veteran. Barrett was production director of the manufacturing-development complex in Luxembourg before becoming production vice-president in June 1976. Virtually all his years with the company had been in the production organization, including posts at the Akron, Gadsden, Topeka, and North Chicago plants.

Stanley J. Mihelick succeeded Barrett as vice-president of domestic tire, retread, and textile manufacturing; for four years Mihelick had been manager of the Gadsden plant, then the world's largest tire manufacturing facility.

At the same time, J. Robert Hicks, president and chief operating officer of Goodyear Canada, returned to Akron as corporate VP of finance. His post in Canada was filled by Albert W. Dunn, managing director of operations in South Africa. Before his South African assignment, Dunn had been VP and general manager of Goodyear Philippines for 10 years.

In line with Goodyear's development of multinational character, all in this series of management moves, except Mercer and Mihelick, had previous experience in international operations. Emphasis on foreign experience would continue during Pilliod's term as chief executive officer.

## GICers Get Corporate Titles

In January 1979, Hicks was elected an executive VP of finance and a member of the board of directors, replacing Bruce M. Robertson who retired after 13 years in that position. In the same month, Edwin M. Humphrey was elected executive VP and appointed president of the general products division. He replaced Richard A. Jay, who continued as vice-chairman and a director.

In line with the continuing growth of international business, which produced sales of $2.8 billion in 1978, two Goodyear International vice-presidents, Joseph C. Graden and Jacques R. Sardas, were elected parent company VPs in December 1978. They were the first GIC officers other than its president to hold corporate vice-presidencies.

Graden, a 28-year veteran, began his foreign service in Mexico, managed plants in the Philippines and Australia, was director of general products at Craigavon, Northern Ireland, and returned to Akron as director of international production. He was named GIC VP for production in 1975. Sardas, a Brazilian, joined Goodyear in Brazil as a salesman. He was sales manager there and in France, became president and general manager of Goodyear-France in 1979, and went to Akron in 1974 as GIC's VP for operations.

The reins of Goodyear's public relations operations also changed hands that month. Robert H. Lane retired as vice-president. Lane was generally credited with making Goodyear blimps the number one corporate symbol in America, was instrumental in building the company's domestic and international auto racing operations, and built one of U.S. industry's most aggressive public relations programs.

William L. Newkirk, since 1972 director of PR under Lane, was appointed corporate director of public relations. A veteran newspaperman, Newkirk had joined Goodyear in 1959 and moved up through the company's PR division, becoming VP in January 1980.

William R. Fair, manager of the Washington and Southern Region PR office, succeeded Newkirk as director; Fair had been in the division 15 years. With these two promotions, Goodyear adhered to its longtime policy of developing executives from within the company — even in the new, sometimes termed exotic, field of public relations.

The man who guided Goodyear through its greatest period of growth ended his official association with the company on April 2, 1979. Russell DeYoung, who had joined the company as a trainee in 1928 and served as board chairman from 1964 to 1974, retired at 70 as chairman of the executive and finance committee of the board of trustees. During his tenure of 10 years as chairman, sales increased from $1,731 million in 1963 to $4,675 million in 1973. In that decade of rapid expansion on all fronts throughout the Goodyear world, net income grew from about $81 million to $176.5 million.

On his retirement, the hard-driving DeYoung gave voice to the philosophy that had guided him, and the company during his years of leadership, in his half century of service.

"One thing we can never do is take our lead over our competitors for granted," he said. "The best way to stay ahead is don't waste a lot of time looking back at your record or that of your competitor's, but devote all of your energies to moving forward, all of the time."

Six years after Chuck Pilliod had assumed the corporate presidency and started the drive to make Goodyear the world's leading radial tire manufacturer, the company could say rightfully that it was the world's largest producer of radial tires. By 1980, it also claimed to be number one in radial tire quality, and the steady increases in its radial tire sales seemed to substantiate the claim.

## Target: Perfection

The road to radial ascendancy was paved with heavy outlays of research and development dollars, continuing expenditures for new and advanced equipment, stepped-up education for the entire development-production work force, and a well-organized plan to achieve a quality and uniformity of product that competition could not match. Whether any other rubber company at that time had the resources, financial and technical, to mount a drive of similar size and scope is doubtful. At least none tried.

Goodyear's attitude in its campaign for supremacy in radials was exemplified by a program in all tire plants called Operation Perfection. Initiated in 1972 under Gerstenmaier's direction, — he was then executive vice-president for production — its basic goal was to measure each plant's capability to meet the required specifications for the tire it produced, and then improve them beyond those specifications.

As management recognized that tire making in the Radial Age was much more a science than ever, it took dead aim on "perfect" radial tire quality and uniformity through Operation Perfection and its many new kinds of electronic testing equipment. An Operation Perfection department was set up in corporate headquarters to work with the development division and the tire factories and to educate service managers in the stores and dealerships on tire quality and uniformity.

The program was rolling smoothly by the late 1970s. Constant evaluation of every piece of factory equipment was routine, and Goodyear engineers were convinced their radials were equal or superior to radials produced by anyone else. In 1978, Stanley J. Mihelick, newly elected vice-president for tire and textile manufacturing, took over direction of Operation Perfection, and the push for perfection was accelerated.

Tire factories were challenged not only to produce better tires than those of rivals, but to make tires superior to each vital characteristic of any other tire. For example, if tests showed tires made by Company A performed best in the wet, tires by Company B were best on expressways, and tires by Company C were most durable, the goal of Operation Perfection was to manufacture a single tire that would outperform and outlast each competitive tire in its own outstanding characteristic.

At the outset, Operation Perfection was a costly operation requiring the scrapping of all production line tires below the program's standards. But, as the program caught on and standards were met consistently in each step of the manufacturing process, the volume of scrapped tires fell rapidly. As Mihelick noted, higher scrap early on meant extremely low scrap later plus new efficiencies because of fewer production problems.

An additional program goal was to establish new targets above and beyond the original equipment specifications set by such exacting customers as the auto manufacturers. Goodyear constantly established and met its own more demanding goals. At the beginning of 1980, with the program just eight years old, less than one percent of Goodyear's radial passenger tires was scrapped for lack of uniformity or failure to meet original equipment requirements. Maybe Goodyear was not producing the perfect tire in mass volume, but it was coming close to it.

Goodyear's "OE or scrap" principle, which required the scrapping of tires that failed to meet requirements for original equipment sales, was a first in American manufacturing. It played an important part in the company's gains in tire market share during the late 1970s and early '80s.

## The Arriva Arrives

A little more than two years after introducing the tremendously successful all-season Tiempo radial, Goodyear hit the market in February 1980 with the all-season Arriva tire. Like Tiempo, the Arriva was molded to the inflated shape, a derivation from the fuel-efficient elliptic tire concept demonstrated in 1977, which provided the lower rolling resistance needed to increase fuel efficiency. Also like Tiempo, Arriva met the industry's definition of a mud and snow tire and rode smoothly on dry pavement.

The first television commercial promoting Arriva appeared nationwide on February 12 in coverage of the inaugural day of the 1980 Winter Olympics at Lake Placid. Arriva TV advertising continued throughout the Olympics, and 23 minutes of commercial time were estimated to have reached 60 million households.

*Testing a new Arriva tire at Goodyear's San Angelo, Tex., proving grounds.*

*The Europa sails gracefully over the Parthenon, Athens, 1979. In a decade characterized by leadership with a truly global perspective, Europa has helped introduce the world's No. 1 tire and rubber company to millions of skygazing fans on the European continent.*

# Chapter XVII

# OUT FRONT AND PULLING AWAY

Despite a recession in the United States and many industrialized nations, a depressed U.S. auto industry, and almost universal inflation that boosted costs of labor and materials, Goodyear zoomed into the '80s at top speed, setting sales and earnings records in both 1980 and 1981.

At the annual meeting of shareholders on April 7, 1980, Chairman Pilliod explained that the company's primary goal was to increase its competitive ability on an international scale. He pointed out that Goodyear facilities in all parts of the world had been updated and expanded "to meet the competitive challenge from any source, be it domestic or foreign."

He reported that in the five preceding years, capital expenditures for plant and equipment had exceeded $1.582 billion and expressed confidence in the effort to move ahead of competition while improving profit ratio and return on investment. One reason for his absolute confidence in this area, he said, was the successful transition to new and more efficient manufacturing facilities and systems, which had given Goodyear the "largest and most sophisticated radial tire production capability in the world and had positioned the company to meet the manufacturing and marketing challenges of the '80s."

The transition had required closure or sale of several outmoded and inefficient facilities. Plants in Los Angeles and Scotland were shut down; so was the Lee tire factory in Conshohocken, Pennsylvania. Plants sold were the P V C resin facility at Plaquemine, Louisiana; a small monofilament factory at Mount Pleasant, West Virginia; and Geneva Wheel, a supplier of small wheels and axles primarily to the industrial equipment industry.

Financial results for the first two years of the '80s seemed to justify Pilliod's optimistic expectations for the new decade. Sales and profits reached new highs in 1980, a year in which no other major U.S. tire and rubber company showed profit gains over 1979, and only one showed any profit at all. Goodyear's sales were $8,444 million, an increase of about $106 million, and profits reached $230.7 million, more than $84 million over the preceding year.

Testifying to the value of geographic diversification, foreign earnings made up 70 percent of the corporate total on sales, amounting to 44 percent of the overall total. Foreign sales of $3.8 billion were greater than the total sales of all but 100 companies in the annual *Fortune* listing of the 500 largest industrial corporations in the United States.

At the annual meeting of shareholders on April 9, 1981, Pilliod said it was possible that international business could represent 60 to 65 percent of total corporate sales by 1985. Reporting that sales in the fourth quarter of 1980 were almost evenly split between the

United States and foreign operations, he explained that the rate of growth in many countries, particularly developing nations, had been more rapid than in the United States.

"Geographical diversification," he said,

> has served the corporation well, in many cases as well or better than we could have realized from a similar emphasis on diversification of our product lines, although we would favor both avenues in the long run. Last year provided an excellent example, as we were able to offset the downturn in the U. S. economy to some extent by increased foreign earnings, which by year's end represented more than 70 percent of our worldwide corporate profits.

## Efficiency and Austerity

Then in 1981, Goodyear increased its market share at home and abroad and became the first company in the tire and rubber industry to top $9 billion in sales, with $9.1529 billion. Net income was at a new high of $260.3 million, about $30 million greater than in 1980, but this time U. S. operations accounted for 65 percent of total net income and 44 percent of total sales.

Just as satisfying to management — and stockholders — was the corporate debt reduction in the first two years of the '80s. Consolidated debt declined by $262 million in 1980 and by $140.2 million in '81. The bulk of major capital expenditures for new and improved plant and equipment had been completed. With that reduction, tight management practices, and rigid inventory controls, total debt was brought down to 40.6 percent of total debt and equity in 1981, from 47.8 percent in 1979, and to 33.1 percent in 1982.

Due to a deepening recession in the United States and abroad, the problems of U. S. automakers, and a continuing flow of tires from foreign countries, particularly Japan, life for Goodyear in those first two years of the '80s was not, however, all cakes and ale. Managers at every level were continually asked to squeeze every inch of fat from their operations; even the smallest expense budgets were scrutinized, and efficiency and austerity were key words throughout the worldwide organization.

If one were to judge from the upsurge of physical fitness activity in and around Goodyear Hall and its big gymnasium, cakes and ale and their softening effects had little appeal for Akron Goodyearites. At dawn, lunch periods, and dusk, the gym was noisy with the padding feet of joggers, the friction and squeals of weight and exercise machines, the slaps on volleyballs, and the smack, smack, smack of basketballs against floor and backboards. At noontime, hordes of people in running suits emerged from the hall fresh and dry, returning a little later bathed in sweat after runs through the headquarters complex neighborhood.

## New Look for Goodyear Hall

A major reason for the rush to barbells, exercise pulleys, jogging tracks, and saunas was the remodeling of Goodyear Hall started in 1976 by Chuck Pilliod, the former footballer, wrestler, and lifelong devotee of physical fitness. The three-phase project included a refurbishing of the exterior, mainly through glare-resistant windows; renovation of the lower three floors; and installation of such things as new lighting, rest rooms, more showers and saunas, additional lockers, a dressing room for women, a new lobby, business library, barber shop, and a wide range of athletic and exercise equipment.

In 1980, an average of 1,200 employees used Goodyear Hall's facilities weekly for exercise and recreation, and 48 company-sponsored clubs scheduled regular meetings there. Sports and exercise programs in the gym area involved more than 1,150 employees. Many participants followed personalized physical fitness programs professionally charted; others worked out in group programs or on their own. Fifty basketball and 46 volleyball teams competed regularly on the gym floor, and nearly 1,200 Goodyearites — both active and retired — participated in league and open bowling on the lanes beneath the gym. Only two years later, 1,800 employees each week were using the athletic-recreational facilities in the hall.

Although Goodyear was one of the best-known companies in the United States and far ahead of its competition in sales, its management was concerned about the possibility that its technological capabilities and industry leadership were not sufficiently appreciated by the general public. Early in 1979, therefore, the company embarked on its first major corporate advertising campaign, designed to complement and reinforce product advertising and to emphasize Goodyear quality, technology, and leadership. It was highly visual and employed both television and magazine media.

Headlines included "Out Front and Pulling Away," "Out Front World-Wide," and "Put the Blimp Behind You." The "blimp behind you" ads pictured motorists, such as a honeymoon couple, a young family, and even nuns, who discover the blimp hovering behind their cars and are reassured that Goodyear, as symbolized by the popular airship, is truly "behind" them.

## Unparalleled Envoys

Entering the '80s, Goodyear was "out front and pulling away," as the exuberant ad claimed. It had consolidated its position as the world's largest producer of radial tires, was gaining larger shares of international and domestic tire markets, was becoming a more important factor in the U.S. aerospace program, and, despite a global recession, continued to establish sales and profit records for the global tire and rubber industry.

Goodyear was the industry's biggest advertiser in the United States and almost certainly in the rest of the world on a global basis. Its four blimps were seen by millions as Goodyear ambassadors of the skies and on TV screens throughout the world. Its racing programs, providing plenty of material for its public relations and advertising campaigns, helped polish its image as a dynamic, technically oriented company — especially among the young.

The U.S. blimps — *Mayflower, Columbia,* and *America* — an accustomed part of the American scene for many years, roamed the country on carefully organized PR schedules supporting charitable and public-service programs, celebrating municipal anniversaries, assisting in ecological studies, helping in major Goodyear promotions, and providing aerial television coverage of national public events. On almost any weekend a Goodyear blimp visited millions of American homes as a participant in TV coverage of big sports events. Such major contests as a world series, a motor race of international interest, a golf or tennis championship, and a key collegiate or professional football game invariably hosted a Goodyear blimp, and the American sports public expected it.

Since 1972, the airship *Europa* had been operating in much the same fashion as her more experienced sister blimps in the United States. Early each spring she departed her home base outside Rome on tours set up in conjunction with Goodyear subsidiaries in the European region, returning just before winter settled over northern and middle Europe. Gradually, *Europa* was becoming a familiar sight throughout Europe. Her travels had taken her as far from Rome as Finland and Sweden to the north, Spain to the west, and Greece to the east.

In 1980, *Europa* visited Belgium, Denmark, England, France, Scotland, Wales, and West Germany. Her schedule included the 1,000th anniversary of the city of Liege, Belgium; the Cologne Cathedral's Centenary; Stuttgart's Canstatter Volksfest; the Paris Auto Show; a three-week summertime visit to the French Riviera and its crowds of vacationers; and TV coverage of major soccer games in England and several big motor races on the Continent.

During the 3-day news blanketing of the July 29, 1981, royal wedding in England of Lady Diana Spencer and Prince Charles and its attendant events, *Europa* assisted in TV's constant coverage. During the pageant preceding the marriage ceremony, the airship operated for 25 hours over 3 days, providing aerial views for the British Independent Television News. The British Parliament gave special permission for *Europa*'s night sign to flash "Loyal Greetings" over London on the wedding night.

## Championship Seasons

As *Europa* built favorable recognition for Goodyear where it had been nearly a stranger a decade or two earlier, so too did the international racing program. Goodyear was the dominant tire supplier for the big international motor races through the '70s and into the '80s and was generally recognized as such.

Since entering the Grand Prix-Formula One circuit in 1965, Goodyear had supplied tires, tire testing, and tire service for 12 of 18 world driving champions.

The company's baptism in Grand Prix racing was on January 1, 1965, when Australian Jack Brabham ran on Goodyear tires in the South African Grand Prix. In the following years, 146 of 235 championship Grand Prix races were won on Goodyears. Dunlop and Firestone, which had dominated the Grand Prix tire scene until the mid-1960s, dropped from the sport in the early 1970s. In 1978,

Michelin entered Grand Prix racing and succeeded in capturing the world Grand Prix tire title in 1979, but Goodyear was on top again in '80, '81, and '82.

Drivers who won world championships on Goodyear tires in the 1965-82 period were Jack Brabham and Alan Jones of Australia; Denis Hulme, New Zealand; Jackie Stewart, Scotland; Emerson Fittipaldi, Brazil; Niki Lauda, Austria; James Hunt, England; Mario Andretti, United States; Nelson Piquet, Brazil; and Keke Rosberg, Finland. All cooperated in Goodyear promotional activity while in competition, and two continued their association with the company long after retirement.

Brabham, now Sir Jack, frequently assists in sales promotions in Australia, and Stewart has been a consultant to Goodyear since retirement after the 1973 season. As the winningest Grand Prix driver in history, the articulate Scot holds a big following among motor-racing fans. He has participated in Goodyear PR activities all over the world, narrated Goodyear-sponsored films, for several years wrote a motoring column distributed worldwide by Goodyear's international PR divisions, and, under the auspices of Goodyear, is a spokesman on motoring safety in visits to many countries. He also is retained by the tire development division as a consultant and tire tester.

*Top, champion racer Gordon Johncock at pit-stop during the 1979 Indianapolis 500. Every Indy 500 winner since 1972 has rolled to victory on Goodyear tires. Right, former world's driving champion Jackie Stewart (left) and Goodyear President Victor Holt, Jr., following announcement that Stewart would ride on Goodyears in the 1971 Formula I Grand Prix. The Scottish driver went on to win that year's championship and, since his retirement in 1973, has continued his Goodyear relationship as PR representative and racing consultant.*

International motorcycle racing rapidly gained popularity in the 1970s, and in 1978 Goodyear became a factor with its sponsorship of the American champion Kenny Roberts of Modesto, California. In 30 major 500cc Grand Prix motorcycle races from 1978 through 1981, Roberts won 14 and gained three world championships.

Goodyear has been the dominant tire supplier on the big-time American motor-racing circuits since the mid-1960s. After the first tentative efforts at stock car tracks following a return to racing in the 1950s, the racing division expanded its scope in the '60s and began to challenge other tire suppliers in most of the more popular championship series.

The company's first Indianapolis 500 victory in the modern era was in 1967 when A. J. Foyt was first under the checkered flag. Since 1972, every Indy 500 championship has been won on Goodyears.

Similar success came in the United States Auto Club (USAC) and Championship Auto Racing Teams (CART) championship series, in which Indy 500 drivers compete. Goodyear tires have finished first in 14 of those championships in the past 18 years, with an unbroken skein from 1973 through 1982.

The other major championships show the same kind of Goodyear dominance: 17 of 25 NASCAR championships since 1958; 14 of 15 Sports Car Club of America (SCCA) Can-Am titles since 1966; 14 of 16 SCCA Trans-Am championships since 1971; and all 12 International Motor Sports Association (IMSA) GT (Grand Touring) championships since 1971.

In the three major categories of Drag Racing Championships sanctioned by the National and International Hot Rod Associations, Goodyear tires were on 43 of the 46 championship vehicles in the period from 1975 to 1982.

Obviously, Goodyear has been gaining promotion benefits from its associations with such American motor racing greats as Foyt, Andretti, Bobby Unser, Gordon Johncock, Johnny Rutherford, Richard Petty, Darrell Waltrip, Cale Yarborough, David Pearson, and others, plus such dragster champions as Don Garlits, Don Prudhomme, and Shirley Muldowney.

## Mercer: Operating Chief

Executive changes made big news within the Goodyear family in the early 1980s. Among the most interesting was a realignment in top management, effective January 1, 1981, in which President Robert E. Mercer became chief operating officer to succeed the retiring vice-chairman, John H. Gerstenmaier, and four group executive vice-presidents were elected. Mercer, with 34 years of service, had been president since May 1978.

Gerstenmaier had joined the company in 1938 as a member of the training squadron and rose through the production ranks in both the tire and industrial products divisions. He was president of the subsidiary Motor Wheel Corp. from 1964 to 1967, when he returned to Akron as Goodyear's director of domestic manufacturing, becoming VP of domestic manufacturing later in the same year. He was elected executive VP for production and a director in 1971 and president in April 1974.

Announcing the group executive vice-presidencies, Goodyear's first, Chairman Pilliod said the offices were established "to strengthen the corporation's organizational structure in accordance with the growing complexity and requirements of our business at home and abroad."

The new group executive VPs were Tom H. Barrett, corporate manufacturing and related services; J. Robert Hicks, finance and planning; Jacques R. Sardas, North American operations; and Ib Thomsen, international operations. Thomsen continued as president of Goodyear International Corporation, a position he had held since 1972.

Barrett joined the company at the Topeka tire plant in 1953 and moved up through the production organization. In 1968, he was awarded a Sloan Fellowship at MIT, earning a Masters Degree in Business Administration. He was production director of the manufacturing and development complex at Luxembourg from 1973 to 1976, when he was elected corporate VP for domestic tire, retread, and textile manufacturing. He was elected executive vice-president, corporate manufacturing, in 1978.

Hicks joined as assistant comptroller in 1962 after 15 years with General Electric and became comptroller in 1964. In 1978, after two-and-a-half years as president and chief executive of the Canadian subsidiary, he returned to Akron as vice-president, finance, of the parent company. In January 1979, he was elected executive VP, finance and planning.

Sardas, a native of Egypt, joined Goodyear in Brazil as a salesman in 1957, rising to sales manager eight years later. He moved to Goodyear-France in 1967 as sales manager and three years later was named president and general manager of that subsidiary. He came to Akron in 1974 as VP of Goodyear International and in 1978 was elected a corporate VP.

Thomsen, a native of Denmark, joined the Goodyear International finance training program in 1952. He served as a trainee in Sweden, then treasurer of Goodyear-India; he became treasurer of Goodyear-Great Britain in 1958. With the British subsidiary, Thomsen was successively financial director and secretary, assistant to the managing director, deputy managing director, managing director, and, in 1968, chairman. He was named a vice-president and director of Goodyear International in Akron in 1971, a year before becoming its president.

The assignment of the four group executive vice-presidents was in keeping with Goodyear's multinational character and need for a global perspective at the top. All four had held important posts outside the United States, and both Sardas and Thomsen gained most of their experience abroad.

The realignment also provided that the presidents of North American Operations (Sardas) and Goodyear International (Thomsen) both would report to the corporate president (Mercer). Previously, both the corporate president and the president of International had reported to the board chairman. As chairman Pilliod explained, with the heads of the domestic and international operations now reporting to the chief operating officer, President Mercer, the multinational balance would be more even.

## Three Key Figures

Three other major assignments became effective on January 1, 1981. F. Vincent Prus, president and chief executive officer of the subsidiary Goodyear Aerospace since June 1981, was elected an executive vice-president of the parent, with responsibility for corporate manufacturing, to replace Barrett. Prus had joined Goodyear in 1949 and transferred to Goodyear Aerospace in 1952, serving there in management positions in the rocket, missile, and electronic divisions, and in 1965 was named production manager of the subsidiary's Arizona division.

*Dye-colored water is spread over a glass plate in preparation for a hydroplaning test at San Angelo, Texas. Goodyear's test operations have been headquartered in San Angelo since 1958 when a multimillion dollar test site was completed there.*

He was named the division's general manager two years later and made production director at Kelly-Springfield in Cumberland in 1971, returning to Akron two years later as director of manufacturing services of the parent company. He was elected corporate VP for general products manufacturing in 1974.

Prus was succeeded as president and chief executive of Goodyear Aerospace by Robert W. Clark, who had been executive vice-president and chief operating officer since June 1980. His first Goodyear assignment, in 1946, was as a product development engineer in Los Angeles. He held positions in aviation products sales in Los Angeles, Wichita, Kansas, and Dayton, Ohio, before coming to Akron in 1967 to take charge of field sales and market planning. After four years as general sales manager for the aviation products division, Clark became director of marketing when the division and Goodyear Aerospace began consolidating their operations in 1974. He was named VP of marketing in '76 and VP for operations in '77.

Douglas F. Hill, a VP of Goodyear International since 1979, was elected executive vice-president, operations, for the international subsidiary and a VP of the parent company. He had joined Goodyear in 1958. In 1959, he began a series of overseas assignments with Goodyear International, including zone manager, headquartered in the Belgian Congo; regional sales manager in Mozambique; zone sales supervisor in Lagos, Nigeria; general manager of Goodyear-Colombia; managing director of Goodyear-Venezuela; and president of Goodyear-Mexico. He returned to Akron as regional director-Europe in 1974 and moved up to VP of Goodyear International in 1979.

Six months earlier, five veterans had been moved up to key posts. Wendell L. Minor was elected executive vice-president of North American subsidiaries, and Frederick J. Kovac, William R. Miller, Frank R. Tully, and Hoyt M. Wells were elected corporate VPs.

Minor had been vice-president North American subsidiaries since 1977. He had joined Goodyear's production squadron in 1948 and then held various assignments in development and sales. In 1965, he became assistant manager of engineered products, then of manufacturers sales in 1967, before his election as a VP of original equipment tire sales in 1972.

Kovac, who became vice-president of tire technology, spent most of his early career, which started in 1956, in tire cord and adhesives development. He was a tire industry pioneer in the development of polyester cord tires. From 1966 to 1975, he was manager of tire reinforcing systems, then served two years as manager of strategic store planning and dealer development. Kovac transferred to Luxembourg in 1977 as director of the Technical Center.

He succeeded Robert B. Knill, whose 38-year career covered virtually every aspect of auto tire engineering. Knill was named manager of compounding in 1956, manager of automotive tire engineering in 1963, assistant director of tire development in 1968, and director of tire development in 1971, three years before his election as vice-president of tire engineering.

Miller joined Goodyear Atomic Corporation at Portsmouth, Ohio, as an engineer in 1953. He moved to Akron in 1962 as an instructor in the training division and became manager of the Training Center School three years later. Subsequent assignments included senior research engineer, technical systems administrator, manager of corporate safety and workmen's compensation, director of equal employment opportunity, and director of governmental personnel relations.

Elected VP of industrial relations, Tully had been director of manufacturing services since 1972. He had joined the company in 1951 and became superintendent of the metal products plant in Akron in 1960. He went to the Motor Wheel subsidiary as director of product engineering in 1965 and was named vice-president of product engineering there in 1967. He transferred to Kelly-Springfield in 1970 as director of production and returned to Akron in 1972 as director of manufacturing services.

Tully succeeded O. M. (Jerry) Sherman, who became a Goodyear employee in 1939 in his native Brazil. Sherman moved up through a series of personnel administration assignments in Akron, returned to Brazil as personnel manager of Goodyear do Brazil, was personnel manager at the Jackson, Michigan, tire plant, manager of administrative engineering in Akron, plant manager at Los Angeles, and corporate director of personnel before becoming a vice-president.

Wells, elected vice-president of general products manufacturing, had for three years been an account executive in Detroit. He had started with Goodyear in 1951 as an engineer at

the Lincoln, Nebraska, belting and hose products factory and became manager of engineering at the North Chicago hose plant five years later. He transferred to the Saint Marys, Ohio, industrial products plant in 1967, becoming manager there two years later. In 1972, he was named VP of general products manufacturing for the subsidiary Goodyear-Canada, a post he held until assignment to Detroit.

## Experience Gained Abroad

Three other executives with international experience were given major assignments in 1981. Westi Hansen was elected an executive VP on May 7. He had been named president of the general products division a week earlier, replacing Edwin M. Humphrey, who would retire later in the year. Before becoming a vice-president of the parent company and assistant to the chairman in 1980, Hansen, a native of Denmark, had been chairman and managing director of Goodyear-Great Britain for three years and VP of the parent for general products

*The durability of Goodyear's Cleartuf resin is dramatically demonstrated as lightweight, shatterproof, polyester soft-drink bottle "pins" made from the resin are "bowled over."*

for a year. After joining the company in 1948, he held various industrial products division posts and was named general manager of the division in 1973.

Robert Milk, vice-president of general products development, was elected VP of product quality and safety to succeed Joseph Hutchinson, who retired early in 1982. Milk started as a laboratory technician in the Niagara Falls, New York, chemical and vinyl resin plant and was named manager of chemical plants engineering in Akron in 1962.

Six years later, he became manager of the chemical plant in Le Havre, France, returning to Akron in 1970 as manager of Goodyear International's chemical division. Five years later, he was appointed director of purchases for the parent company, holding that position until 1980 when he was elected VP of general products development.

Hutchinson, with a mechanical engineering degree from Georgia Tech and postgraduate work at the U. S. Naval Academy, joined the company in 1947 as a production supervisory trainee and in 1949 participated in the Sloan Fellowship program at MIT. He held several engineering posts before appointment in 1960 as manager of automobile tire engineering. He became director of product quality and safety in 1969 and in 1972 became the first vice-president for product quality and safety.

Milk was succeeded as VP of general products development by John Fiedler, managing director in Malaysia. Fiedler joined the chemical division in 1964 and after a series of sales engineering posts was named product manager for fiber resins in 1969 and polyester marketing manager in 1972. He was assigned to special corporate projects in 1976 as manager of products and materials coordination, and during the year preceding his 1980 assignment to Malaysia, he was assistant to the board chairman.

A few months earlier, in August, a new title had been added to the corporate list, VP responsible for environmental safety and health programs. It reflected the company's concern for those areas of corporate responsibility and its response to the increasing burden of administrative work required by the federal government's growing regulatory effect on industry.

This post was filled by Dr. Robert M. Hehir, a former U. S. government executive who had joined Goodyear in 1980 after 20 years as a toxicologist with federal agencies. He had been a branch chief of the Food and Drug Administration's Hazardous Substances Laboratory, then director of the Bureau of Biomedical Service of the Consumer Products Safety Commission, its deputy associate director of health sciences, and finally its director of all laboratories.

The beginning of 1982, the last full year of Chairman Pilliod's tenure, held few encouraging auguries for U. S. business and industry. Interest rates were at extraordinary highs. Inflation, although starting to decline, was still at worrisome levels. Capital investment was sluggish and productivity low. An overvalued dollar inhibited U. S. foreign trade, and unemployment was rising steadily.

Overseas the picture was no brighter. Europe slid deeper into recession, Japan felt stronger resistance to its national export program, and many developing nations staggered under massive debt burdens.

But the world's largest tire and rubber company looked ahead with confidence, a quality it had demonstrated consistently for 84 years. Goodyear had sound reason for optimism. Its leadership position in the world tire industry was solid, perhaps stronger than ever. Most of the capital investment for conversion to radial tires and expansion of its radial capability was complete, its industrial products business healthy.

The commitment to advanced technology and technical excellence was solid and companywide. Corporate sales and profits had reached record highs in 1981. Earnings in 1982 were $264.8 million, another record, but sales dropped to $8.69 billion from the record $9.15 billion of 1981.

As the Pilliod era neared an end, the company's chief executive since 1974 had accomplished his major goals of consolidating Goodyear's industry leadership, moving it to the top in radial tire sales, and orienting the entire organization to innovative technology.

The worldwide recession had a negative effect on Goodyear's international business. Operations in Mexico, where $107 million was invested in 1980 for Mexico City plant expansions over four years, were slowed by an unexpected recession.

Continuing weakness in the economies of Europe and the United Kingdom hurt the tire and industrial products business there. Even the economies of Southeast Asia, which had led the world in growth since the late 1970s, were slowing. As a result, Goodyear's international sales and earnings dropped from 1981, 11.2 percent for sales and 65.8 percent for earnings.

Management's attitude in 1982 regarding its foreign operations was one of concern, but not dismay. Only two years earlier foreign sales and earnings had reached highs as earnings contributed 70 percent of the corporate levels. The Goodyear International veterans in top management — of whom there were many — had been through numerous ups and downs in international markets.

- Loss of the Indonesian operations in World War II and again in 1965
- Expropriation of the Cuban factory in 1960
- Massive inflation and currency devaluations in South America
- The Peron years in Argentina
- Lean times in India where government relations often were difficult
- The growth and sometimes retraction of the welfare state in Great Britain and Australia
- The advance and relapse of certain African economies
- Government instability in many countries at one time or another

These long-time international managers, in fact, had seen or experienced the gamut of economic and social change around the world for more than half a century.

The knowledge that Goodyear's position in the global marketplace was firmer, its diversification wider, and its resources greater than those of its competitors undoubtedly gave management a long-range perspective that was not shaken unduly by socioeconomic disruptions on the international scene.

## Changing Face of Akron

In Akron, the Goodyear scene underwent some highly visible changes. The sign over the bank building adjacent to Goodyear Hall no longer displayed "The Goodyear Bank." Arrangements had been started in 1981 for the bank's sale to the National City Bank of Cleveland and were completed in 1982. At the time of sale, the Goodyear Bank had assets of $285 million, 12 offices providing a full range of banking services, and 175 employees; all personnel were retained by National City.

The bank had been founded in 1933 to provide financial services to Goodyear enterprises and employees, but in the 1960s and 1970s addressed a broader clientele, expanding its branch system and services to the general public.

A more dramatic change in the appearance of the headquarters complex came with the demolition of several Plant I buildings that dated back to the company's early days.

The razing or emptying of tire industry factories in Akron was commonplace in the late 1970s and early 1980s. By 1982, no volume tire manufacturing took place in the city. Antiquated two- and three-story factory buildings unsuitable for expansive renovation; high labor rates; an industrial relations climate less salutory than in other parts of the country; the lure of tax advantages in the industrializing South and Southwest: these combined to change the Rubber Capital of the World for manufacturing to the Rubber Capital of the World for administration and research.

As the tire and rubber industry spread its manufacturing operations throughout the country, its manufacturing employment in Akron declined steadily from 1950. The total Akron work force of Akron's Big Four — Goodyear, Firestone, Goodrich, and General — dropped from about 51,000 in 1950 to about 37,000 in 1970, to 23,000 in 1980, and to 21,000 at the start of 1982 and fewer than 20,000 at its end. Goodyear-Akron employment decreased from 13,500 in 1950 to 12,047 at the end of 1982, including 4,065 employees at GAC.

Akron itself survived its loss of manufacturing employment in fairly good shape. Despite plant closings, rubber industry layoffs, and the recession, population and employment remained stable. Jobs lost in manufacturing were absorbed by service businesses, retail stores, high technology operations, and increasing administrative and research-development operations in the Big Four's headquarters complexes.

The "silk shirt" days, though, that followed World War I when thousands flocked to Akron for rubber industry jobs were gone forever, as were the halcyon times of post-World War II when the city's rubber industry employment was at its ascendancy.

## Well-Trained Replacements

Several names on top-management officer doors were changed in 1982, but as it had over the years, Goodyear's essential character and basic philosophy remained the same. As old hands moved to retirement, others — often their number one assistants — moved in, trained and ready. This was exemplified in 1982 by changes in the top financial echelon. In April, G. Alexander Sampson, vice-president and comptroller, retired after 38 years with the company. He was succeeded by James R. Glass, assistant to the group executive VP of financial planning.

In October, Glass was elected group executive vice-president of finance and planning and a director. He succeeded J. Robert Hicks, who took early retirement as of December 31, and James E. MacDonald succeeded Glass as vice-president. Both Glass and MacDonald had extensive careers with Goodyear International.

Glass had become a Goodyearite in 1964 as comptroller of the Motor Wheel subsidiary and was named to the same post for Goodyear International in 1965. He later became VP and treasurer for Goodyear International and assistant treasurer of the parent. Beginning in 1972, Glass spent seven years in overseas assignments, including president of Goodyear-Japan and president-director of Goodyear do Brazil before being elected president of Motor Wheel in 1979. He returned to Akron in January 1982 as assistant to Hicks, then VP and comptroller.

MacDonald joined the company in 1950 in Boston and went into international operations in 1958. He was assigned to Belgium as regional credit manager for Europe in 1961. He became credit manager for Goodyear International in 1963 and went to Japan a year later as finance director of the Japanese subsidiary, subsequently holding the same post in South Africa, Turkey, and Australia. MacDonald went back to Goodyear-Japan as president in 1975 and returned to Goodyear International headquarters a year later as comptroller.

Another veteran international finance expert joined Glass and MacDonald at the top of the company's financial management pyramid in January 1983. Thomas E. Wightman, financial vice-president and treasurer of Goodyear International and assistant treasurer of the parent, was elected a corporate VP. Wightman joined the company as an international trainee in 1950 and was assigned to Goodyear-Argentina in '51.

In 1953, he was appointed assistant treasurer of Goodyear-Mexico and moved to Peru as secretary-treasurer seven years later. He came back to Akron in 1963 as finance manager for the Asia-Africa region and later that year went to South Africa as finance director. He returned to Akron in 1967 as assistant comptroller of Goodyear International and moved up to comptroller in 1968.

Wightman was named assistant treasurer of the parent company and VP and treasurer of the international subsidiary in 1972.

For the first time in Goodyear's history, three internationally trained financial men — Glass, MacDonald, and Wightman — had key posts in the company's finance division. The other top finance position, VP and treasurer, had been held since 1977 by Donald R. Kronenberger, whose experience had been solely in the U.S.

An important change in sales management took place in '82 when Charles A. Bethel, VP of original equipment sales and a 36-year veteran, retired and was succeeded by James A. Bailey, director of marketing for original equipment tires. Bailey had 31 years of service.

The end of Goodyear's 84th year, in which earnings reached a record high, brought a new hand to the helm and the Pilliod era to its end at high levels of achievement.

## End of the Pilliod Era

On December 9, 1982, at a crowded management meeting in the Goodyear Theater across Market Street from corporate headquarters, Charles J. Pilliod, Jr., announced he would step down as chief executive officer, effective December 31, and that the directors had elected Robert E. Mercer, president since 1978, vice-chairman and CEO effective January 1, 1983. Pilliod would continue as chairman of the board and as chairman of its finance and executive committee.

Ten months short of his 65th birthday, Pilliod explained he was relinquishing the top management post before mandatory retirement to ensure a smooth transition and give the new top management team time to "flex its muscles." He also expressed his gratitude to the team he had led since 1974.

"It's no secret that what has made Goodyear great and kept us in front is our people, from those on the factory floor to those on mahogany row," he said.

> For many years, all kinds of outside observers — from Wall Street analysts to financial writers to our competitors — have recognized the high quality of our management ranks and have gone on record to that effect. And when such plaudits are given to our management, they are not referring just to the guy in the chairman's seat, or the president, or the executive vice-presidents. Invariably it is the depth of our management that is recognized, applauded, and even envied. Our whole management team shares and deserves that credit, and in all sincerity nothing in my life has been more rewarding for me than being the leader of that team.

As Mercer moved up to the top post, he was succeeded as president and chief operating officer by Tom H. Barrett, who had been group executive VP for corporate manufacturing and related services since 1980. Stanley J. Mihelick, VP for tire and textile manufacturing since 1978, succeeded Barrett. Mihelick was replaced by Richard A. Davies, manager of the Union City, Tennessee, radial tire plant since 1975. Davis was well-remembered by company sports fans as one of three Goodyear Wingfoot stars who had helped the U.S. team win the 1964 Olympic basketball championship at Tokyo.

The Mercer succession, expected by insiders and rubber industry observers since his election

as president in 1978, stimulated speculation about a change in Goodyear's management style. Pilliod exercised a personal, driving, hands-on leadership that sometimes resulted in his direct involvement in such activities as tire design, middle-level executive changes, and even the improvement of advertising copy. Mercer, more retiring, was noted for a likable personality, an analytical mind, and a rapierlike wit.

## Firmness with a Smile

Pilliod might be abrasive to emphasize a point or get a program moving; Mercer usually takes a smoother, lower-keyed approach. As one Goodyear executive said, "Bob can be just as firm as Chuck, but with a smile."

In a *Wall Street Journal* interview, Mercer said: "My style isn't his style. If I'm chief executive, people will be required to exert more leadership in their areas of expertise, although the company won't be run by committee."

Despite the differences in their personalities, Mercer and Pilliod have many similarities. Both have devoted their business careers to Goodyear; they know the business from the ground up; they have been successful in a wide variety of middle management and executive positions; they have shared a global perspective on the company's business; and both not only are respected, but are well-liked throughout the company and the tire and rubber industry.

An indication of the Mercer wit and his friendly relationship with the man he would

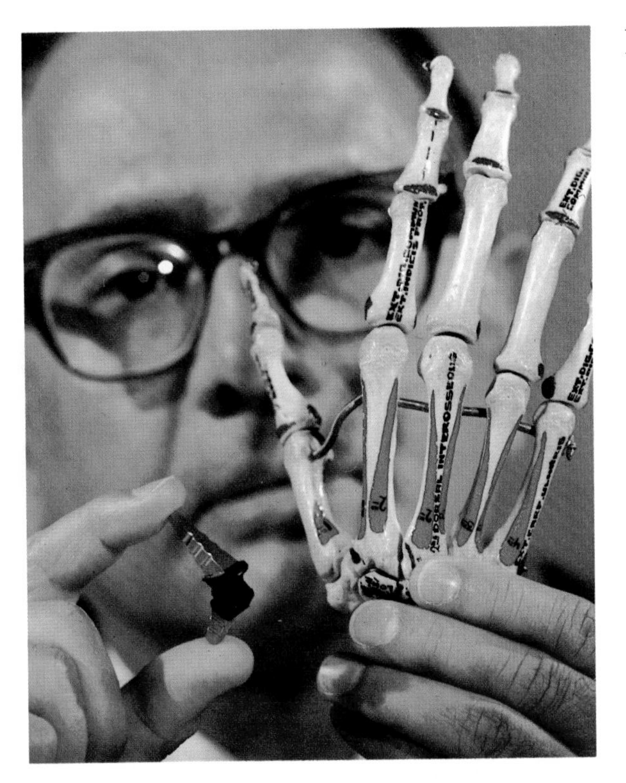

*From small rubber goods and carriage tires to a wide variety of product lines. Opposite, steel-reinforced suction discharge hose at a Louisiana docking facility. Above, a man-made joint that holds promise for victims of diseased or damaged hands. Made of titanium and highly flexible rubber Goodyear originally developed for use in automobile tires, the replacement body part was designed by Lord Corp., Erie, Pennsylvania.*

succeed as chairman is a plaque that hung in the office he occupied as president — next door to the chairman's. It read: "The buck pauses here ... but stops next door."

After his move upward, Mercer told the *New York Times*: "There will be no right turn on the rudder. Mr. Pilliod and I have been working together for about a decade, and we see eye to eye on where the corporation should be headed. The organization is not going to change in the foreseeable future."

Barrett, Goodyear's 14th president, is more reserved than both Pilliod and Mercer, but intensely competitive and a "let's get it done" leader in the positive, charging style of the company's 4th chairman and 9th president, Russell DeYoung. And, like DeYoung, his ascension to top rank was through the production forces, with brief stops in engineering and development.

## Leadership Position

The new team took over at a high point in the company's progress. With the global and U.S. economies in deep recession, the U.S. auto industry still struggling for a return to normalcy, and most major tire and rubber companies reporting declines or losses for 1982, Goodyear established record corporate earnings of $264.8 million while reducing its debt to the lowest level in 17 years, 33.1 percent of debt and shareholder equity.

Offsetting a decline in foreign sales and earnings, U.S. earnings increased 38.2 percent over 1981 to a record of $233.6 million, despite a slight drop in sales.

Observers and analysts of the tire and rubber industry were quick to credit Goodyear's outgoing chief for the company's irrefutable leadership in the industry and its robust economic health as contrasted with its major rivals in the United States and Europe. In early January 1983, the *New York Times* published a front-page story on Pilliod: "*Driving Ahead* — Chief's Style and Ideas Help to Keep Goodyear No. 1 in the Radial Age."

It opened with a brief account of the now-famous story — at least within Goodyear — of Pilliod's abrupt dismissal of his fellow executives' argument that radial tires never would take over the U.S. market and his individual decision to produce 100,000 radials a day.

The *Times* story described Goodyear as:

> a rare example of an old-line smokestack company that has pulled ahead of its domestic rivals and is successfully meeting growing foreign competition. Goodyear is the world's No. 1 tire company. In the U.S., it has opened a wide lead over Firestone Tire & Rubber Co., which was once a strong No. 2 world-wide but now is No. 3 and is shrinking its tire operations.
>
> Goodyear also has overcome the early lead of Michelin Tire Corp. in radial tires and helped blunt Michelin's expansion in the U.S. With its U.S. market share nearly stagnant in recent years, the French concern, now No. 2 world-wide, has indefinitely postponed plans for three new plants in Texas.

Committed to continuing leadership of the tire industry, Goodyear stood securely at its head as it completed 84 years of business. But management was wary of overreliance on an automotive industry that had demonstrated almost tragically its cyclical nature in the late 1970s, at least in the United States.

Completing the bulk of expenditures for the changeover to radials, finishing expansion and modernization of tire production capacity on a global basis, and reducing its debt by a tremendous amount, the company was ready to diversify — preferably in areas unrelated to the auto industry. Accordingly, a diversification task force was formed in early 1982 and given a mandate to prospect for new investment opportunities.

From small rubber goods and carriage tires, Goodyear had expanded into many other product lines and production-marketing activities over the years.

- Auto, truck, bicycle, airplane, agricultural, and earth-mover tires
- Wheels and brakes
- Industrial products
- Engineered products
- Aerospace
- Uranium enrichment
- Computer technology
- Chemicals
- Shoe products
- Films and flooring
- Graphics products
- Lighter-than-air aviation
- Cotton farming
- Coal mining
- Even a resort business at the Wigwam in Arizona

Although it had been the world's largest tire company for 66 years, its 1982 nontire sales — amounting to about 30 percent of the corporate total — were on a level with the sales of the 150th largest industrial corporation in the U.S.

Goodyear's sales had grown from $508,597 in 1899, the first full year of operation, to $8.69 billion in 1982; its employees from about 20 in the fall of 1898 to more than 131,000; and its production facilities from the 3 small buildings of the former strawboard factory on the banks of the Little Cuyahoga River in Akron to 53 plants in the United States and 47 plants in 27 foreign countries.

Now, approaching its 84th birthday, Goodyear was ready to explore new frontiers.

*Although diversification had become a major business trend in the 1970s, the company's policy, as enunciated by Pilliod upon his taking over as chief executive officer, was to firm up the base business — tires, industrial products, and aerospace — before considering moves into new fields. Approaching that point in 1982, and with a debt equity ratio nearing the target level of 35 percent, the initial steps toward diversification were taken. Dennis Rich, GIC's European finance director, was brought to Akron as director of corporate business strategy and analysis for the express purpose of seeking the most likely routes. The effort was culminated on February 8, 1983, when Goodyear's directors unanimously approved an agreement providing for the merger of Celeron Corp. of Lafayette, Louisiana, with a wholly owned subsidiary of Goodyear. Celeron operates natural gas transmission systems, principally in Louisiana, and is also involved in a wide range of related oil and natural gas activities.*

*It also engages in onshore exploration and production programs in 14 states. In 1982, Celeron earned $65 million on revenues of $942.8 million. The transaction will involve a Goodyear investment of about $825 million.*

*Chairman Pilliod explained that Goodyear's diversification goal was a merger with a friendly company whose management wanted "to come with us," a well-run company with a steady business or a business that would be countercyclical to Goodyear's and a low debt equity ratio. If the move were in the energy field, a business less risky than drilling would be preferable — a pipeline company, for example. Goodyear looked for a company that would not bring a heavy debt burden, one that could be merged through a stock exchange and that would not require Goodyear to borrow.*

*Celeron seemed to fill the bill. Its management was regarded as the best in the pipeline business, it liked the idea of a merger with Goodyear, and its debt equity ratio was 30 percent. So, as Pilliod pointed out, "we would be financially stronger than before and would have a positive cash flow to help us move ahead in our established business areas."*

*In announcing the merger, Goodyear reported it was in line with corporate objectives of maintaining its position as the world's number one tire manufacturer while reducing dependency on the automotive field.*

*Thus, Goodyear is now positioned to consider expansion as opportunities occur in four major areas: tires and tubes, general products, defense and high technology through Aerospace, and energy.*

# Epilogue

The Goodyear archives are tucked away on the fourth floor of 62-year-old Goodyear Hall, near the east end of a 200-yard corridor. At the corridor's other end is the company's museum, the World of Rubber, open to the public on working days.

Visitors to the World of Rubber — and on most days there are many — can see and be told about mementos of the Goodyear past.

- A single tube tire of 1898
- A fragment of an early rubber plantation
- Replicas of the tiny 1971 moon tire and the huge 1940 terra tires made for Admiral Byrd's Antarctic snow cruiser
- The race car that won the Indianapolis 500 in 1972
- A fuselage of one of the 4,008 *Corsair* fighter planes produced for the Navy during World War II
- A working model of an artificial heart with a rubber diaphragm
- Even a voluble Charles Goodyear recounting how he discovered the vulcanization of rubber in 1839

The archives are about one twentieth as large as the museum, musty, and always quiet. In logical connotation of their role as the repository of bits and pieces of the corporate past, they have a slight air of disarray.

Models of blimps and dirigibles, plaques, medallions, framed certificates, and a few antique promotion pieces are perched in no particular order atop metal filing cabinets and open shelves stuffed with books and photo albums. For the occasional visitor, the archives stimulate a feeling not unlike that of an attic in an old family house. They sit in sober quietude, waiting for someone to disturb the silence and the symbolic dust in quest of a sign or an echo of the past.

A fresh scent of change is lacking in the archives. The historical material lies inert in the files, books, and albums that are passive guardians of the past akin to the molding boxes and suitcases in a family attic. The archives come to life only when a cabinet, a drawer, or a photo album is opened by a seeker of some historical fact, or some hint of the Goodyear that was, or some image of the people who made it what it was and is.

The archives belong to them, the Goodyearites of yesterday. They could tell us, much better than the archives, all the stories that put together could describe the long journey from 1898 to today.

Neither the loaded files of the archives nor the crowded photo albums can provide a comprehensive account or a panorama of Goodyear's history. But just a hasty look, like a quick visit to the family attic, will uncover some of the spirit of the Goodyear clan, the spirit that made Goodyear. That spirit pervaded the small strawboard factory on the Little Cuyahoga in late 1898 and has remained with the company, spreading as it grew.

It is hoped this condensed history, based mostly on what the archives have disclosed and what veteran members of the clan have said, has captured the essence of the Goodyear spirit. The spirit really is the whole story.

# Bibliography


Allen, Hugh. *Goodyear Aircraft.* Cleveland: Corday & Gross Co., 1947.

— — —. *The House of Goodyear.* Cleveland: Corday & Gross Co., 1949.

— — —. *Rubber's Home Town.* New York: Stratford House, 1949.

— — —. *The Story of the Airship.* Chicago: Lakeside Press, R.R. Donnelley & Sons, 1947.

Beasley, Norman. *Men Working.* New York: Harper & Brothers, 1931.

Beaubois, Henry. *Airships, Yesterday, Today and Tomorrow.* New York: Two Continents Publishing Group, 1976.

Bloemker, Al. *500 Miles to Go.* New York: Coward-McCann, 1961.

Botting, Douglas. *The Giant Airships.* Alexandria, Va.: Time-Life Books, 1980.

Grismer, Karl H. *Akron and Summit County.* Akron, Oh.: Summit County Historical Society, 1952.

Hall, George; Wolman, Baron; and Larson, George. *The Blimp Book.* Mill Valley, Calif.: Squarebooks, 1977.

Hansen, Zeno. *The Goodyear Airships.* Bloomington, Ill.: Airship International Press, 1977.

Klippert, W. E. "Reflections of a Rubber Planter." Peninsula, Oh., 1971.

Kovac, F. J. *Tire Technology.* Akron, Oh.: Goodyear Tire & Rubber Co., 1978.

Kovac, F. J., and Kersker, T. M. "The Development of the Polyester Tire." *Textile Research Journal* 34 (1964): 69-79.

Kovac, F. J., and McMillen, C. R. Polyester Tires, Success Story of Planned R&D. Paper read at 48th Conference of the Chemical Institute of Canada, 2 June 1965, Montreal, Canada. Mimeographed.

Litchfield, P. W. *The Industrial Republic.* Cleveland: Burrows Brothers Co., 1946.

— — —. *Industrial Voyage.* Garden City, N.Y.: Doubleday & Co., 1954.

Manchester, William. *The Glory and the Dream.* Boston-Toronto: Little, Brown & Co., 1973.

Roberts, Harold S. *The Rubber Workers.* New York: Harper & Brothers, 1944.

Schetter, Clyde. "History of Goodyear, 1898-1967." Unpublished. Akron, Oh. Goodyear Archives.

Shilts, W. D. "History of Goodyear, 1898-1927." Unpublished. Akron, Oh. Goodyear Archives.

Smith, Howard R. *Economic History of the United States.* Ronald Press, 1954.

Smith, Richard K. *The Airships Akron & Macon.* Annapolis: United States Naval Institute, 1966.

Wilson, Charles M. *Trees & Test Tubes.* New York: Henry Holt & Co., 1943.

Wright, Chester Whitney. *Economic History of the United States.* New York: McGraw-Hill, 1949.

Various issues of *GIC Newsletter, Orbit, Triangle,* and *Wingfoot Clan,* Goodyear Tire & Rubber Co. internal publications.

# Index